Antonio da Costa Diakos
Blandina Viana
Fabiana Silva

Efficiency of Apis mellifera L. hive densification and management

Antonio da Costa Diakos
Blandina Viana
Fabiana Silva

Efficiency of Apis mellifera L. hive densification and management

Effect of Apis mellifera Supplementation and Hive Management on Seed Production in Malus domestica Bork

ScienciaScripts

Imprint

Any brand names and product names mentioned in this book are subject to trademark, brand or patent protection and are trademarks or registered trademarks of their respective holders. The use of brand names, product names, common names, trade names, product descriptions etc. even without a particular marking in this work is in no way to be construed to mean that such names may be regarded as unrestricted in respect of trademark and brand protection legislation and could thus be used by anyone.

Cover image: www.ingimage.com

This book is a translation from the original published under ISBN 978-613-9-63854-3.

Publisher:
Sciencia Scripts
is a trademark of
Dodo Books Indian Ocean Ltd. and OmniScriptum S.R.L publishing group

120 High Road, East Finchley, London, N2 9ED, United Kingdom
Str. Armeneasca 28/1, office 1, Chisinau MD-2012, Republic of Moldova, Europe
Printed at: see last page
ISBN: 978-620-7-70266-4

To Chiara who cried

To Melody who doesn't know her father

To my parents Antonio and Catarina for transmitting values of respect and conservation of nature

My sisters and brothers for their help and patience

To my friends Daniel, Ivaldo, Kito and Tony for encouraging me to fight for my ideals.

We cannot enter modernity with the current burden of prejudice. We need to take our shoes off at the door of modernity. There are seven dirty shoes that we need to leave on the doorstep of new times:

- First shoe:
- The idea that others are always to blame.
- Second shoe:
- The idea that success doesn't come from work.
- Third shoe:
- The prejudice that anyone who criticises is an enemy.
- Fourth shoe:
- The idea that changing words changes reality.
- Fifth shoe:
- The shame of being poor and the cult of appearances.
- Sixth shoe:
- Passivity in the face of injustice.
- Seventh Shoe:
- The idea that to be modern we have to imitate others.

Mia Couto

Thank you

To God for providing this opportunity to live together, learn and carry out this work;

To Profª Dr Blandina Felipe Viana, for her guidance, understanding, advice, patience and help in difficult times. Thank you very much for everything;

To Profª Di* Fabiana de Oliveira, for her guidance, advice, suggestions and criticism throughout the process of carrying out the project and writing the dissertation. Thank you for everything;

To Profª Drª Favizia Freitas, MSc Thiago Mahlmann and the team at the Insect Systematics Laboratory who helped with the identification of the bees;

To Professors Dr Guido Castagnino and Dr³ Katia Gramacho for their guidance and valuable lessons on inspection procedures and the management of *Apis mellifera* bee hives.

To my friend Jeferson Coutinho, for his patience in teaching, for his partnership, support, suggestions, help and useful advice, without which this work would not have been possible. Thank you for everything;

To my friends Eduardo Moreira, Juliana Hipólito, Lady Catalina, Rafaela Lourena and Uiré Pena for their valuable contributions during the realisation of this project;

To my colleagues at LABEA, for their welcome, collaboration, companionship and contribution to the realisation of this project;

To the beekeeper Claudinei, for supplying the *Apis mellifera* hives used to pollinate the apple trees and for his advice on management;

To the Federal University of Bahia (Institute of Biology) for the opportunity to study for a master's degree in Ecology and Biomonitoring;

To GEF/FAO/INEP/Brazilian Fund for Biodiversity (FUNBIO) for the master's scholarship that made this course possible;

To CNPq and GEF/FAO/INEP/FUNBIO for their financial support in carrying out this research.

To Bagisa do Brasil and its employees for their help in carrying out this project and during the experiment.

Index

GENERAL INTRODUCTION

The apple tree *(Malus domestica* Borkh) originated in Central Asia and has around 7500 varieties (Keogh *et aL,* 2008). It is one of the most widely cultivated plants in the world, including in Brazil, where advances in planting technology related to trellising and pruning, artificial overcoming of dormancy, harvest management, chemical thinning, pollination, phytosanitary control, conservation and storage of the fruit have enabled the expansion of production centres (Fioranço, 2009; Petri *et al.,* 2011).

In the north-east, apple trees of the Eva and Princesa varieties are grown in the state of Bahia, in the Chapada Diamantina and, recently, in experimental areas in the São Francisco sub-middle river region, where the aim is to develop a management system to make apple production viable in the months of October to January, a period of low supply of fresh fruit on the national market. In the São Francisco region, the Eva, Princesa, Condessa, Daiane and Gala varieties are grown (Lopes and Oliveira, 2010). In the Chapada Diamantina, the apple orchard covers 43 hectares and is located in the municipality of Ibicoara - BA, where the climatic conditions are favourable for growing this fruit. The apple trees were introduced to the region between the end of 2005 and the beginning of 2006 and the first fruit began to be harvested at the end of 2007 (Anjos and Oliveira, 2008).

The apple cultivation system requires the intercropping of a commercial variety and a pollinator variety in order to provide viable and compatible pollen, since the varieties are self-incompatible (Free, 1993). Self-incompatibility is the inability of a hermaphrodite plant to produce zygotes after self-pollination (Richards 1997). It occurs due to the inhibition of pollen grain germination or pollen tube growth by the stigma. Fruiting therefore depends on successful pollination, as well as the intrinsic characteristics of the cultivar, which will determine the quality and quantity of fruit produced.

This stage of the reproductive process requires the services of the pollinator, which is able to carry out cross-pollination between compatible varieties, guaranteeing the fertilisation of the ovules and, consequently, fruit set. Thus, the efficient use of bees for apple pollination can result in an increase in both fruit quantity and quality. In the end, proper pollination by bees will ensure adequate seed formation and reduce the incidence of deformed apples, which in turn results in success for the farmer (Delaney and Tarpy, 2008).

The honeybee *Apis mellifera* is considered to be the visitor responsible for around 98% of floral visits in apple trees (Joshi and Joshi, 2010), and is recognised as the most important pollinating insect in agricultural areas (Todd and Mcgregor, 1960). According to Stern et *al.* (2007), A *mellifera* is important for improving cross-pollination in apple orchards. In their work on the management of

Apis mellifera hives, these authors found a significant positive correlation between the number of bees and the percentage of fruit produced per tree in pear and apple orchards, which would result in an increase in production. A *mellifera* hives are introduced during the flowering period to increase the transfer of pollen between the different cultivars of apple trees (Free, 1993). The number of hives required per unit area of an orchard is highly variable and depends on various factors, such as the proportion of the area cultivated, the proportion of natural habitat, plant richness, the gradient of isolation from natural areas, the type of management, the level of toxicity in the area due to the use of insecticides and the abundance of the A *mellifera* bee (Kremen *et al.*, 2004). However, there must be enough bees to visit a sufficient number of flowers so that seed production is maximised (Ferreira *et al.*, 2006).

Different factors can influence pollination in agricultural crops; low bee population density, for example, negatively affects the transfer of viable pollen to receptive flowers, resulting in a pollination deficit. Weather conditions, the spraying of chemical substances, human traffic and the presence of machinery, as well as other orchard management practices can also significantly affect the foraging activities of bees, thus influencing pollination success (Delaney and Tarpy, 2008).

In apple blossoms, for complete fertilisation to occur after pollination, 6 to 7 ovules must be fertilised. Failure to reach this threshold can result in morphological and physical deformation of the fruit, reduced production, the formation of small fruit and a reduction in the amount of calcium with consequences during conservation (Delaney and Tarpy, 2008).

There is known evidence of pollination deficits in several plant species in South Africa, such as *Brassica napis* var. *Oleifera* (rapeseed), *Cynara scolymus* (artichoke), *Eriobotrya japonica* (medlar), *Fragaria x ananassa* (strawberry), *Helianthus annus* (sunflower), *Persea americana*

(avocado), *Rubus* spp. (black berry), *Sinapis alba* (white mustard), *Trifolium spp.* (clover), *Vicia spp.* (pea), *Vicia faba (broad* bean) where the quantity of seeds, fruit production or quality increased due to the addition of *Apis mellifera* hives during flowering (Donaldson, 2002).

In Brazil, knowledge about the pollination of agricultural crops has advanced, with consistent data on pollination requirements in castor bean (Carvalho, 2005), açaí (Venturieri, 2008), cashew (Freitas, 2006), Brazil nut (Maués, 2002), mangaba (Oliveira and Schilindwein, 2005) and yellow passion fruit (Viana *et al.*, 2012). These crops are important for the Brazilian economy, both for export and to satisfy the demands of the domestic market, and these may be the reasons why most of the available studies focus on these crops (Imperatriz-Fonseca, 2004).

This study conducted in the Chapada Diamantina was stimulated by observations previously made which suggested that there was a pollination deficit in the Eva orchard (Silva and Viana 2008, unpublished data). This dissertation presents the results of the studies carried out in 2008, 2010 and 2011, the main objectives of which are discussed in two chapters. In the first chapter, we look at the morphological and functional description of the flowers, duration and intensity of flowering and floral visitors of the Eva and Princesa varieties. We assessed the existence of a qualitative pollination deficit in commercial apple orchards under the local A. *mellifera* management conditions (5 hives/ha) usually employed by the producer. To do this, we compared the number of seeds produced per fruit in the manual cross-pollination and natural pollination treatments in the Chapada Diamantina region of Bahia. In the second chapter, we investigated the effect of densification with 7, 9 and 11 *Apis mellifera* hives/ha, with and without pollen collectors, on the density of visits and seed production in the fruit. From this study we hope to determine the number of hives and the most appropriate management to increase productivity in the orchard.

REFERENCES

ANJOS, A. P.; OLIVEIRA, J. M. Frutas da Bahia: desempenho e perspectivas. v.8, n.2, nov. 2008. Available at: <http:// www.seagri.ba.gov/pdf/2-agrossintese v.8, n.2.pdf. Accessed on: 2 August 2012.

CARVALHO, B. C. L. Manual do cultivo da mamona. Salvador: EBDA 2005, p.65.

DELANEY, D.; TARPY, D. The role of honey bee in apples pollination. North Carolina State University and North Carolina A&T State University Commit. Apiculture Programme, Beekeeping Note 3.03, 2008. Available at: <http;//www.cals.ncsu.edu/.../apiculture/...3.03%20cop...Acesso on: 3 Aug.2012.

DONALDSON, J. S. Pollination in agricultural landscapes, a south African perspective.In: Kevan & Imperatriz Fonseca VL (eds) - Pollinating Bees - The Conservation Link Between Agriculture and Nature - Ministry of Environment/Brasília p. 97 - 104. Available at: <http;//www. Webbee.org.br. Accessed on: 14 August 2012.

FERREIRA BAPTISTA, J. G.; BESSA BATISTA, E. R.; PACHECO DE MEDEIROS, C. A. Effect of bee pollination and polliniser proportion on apple productivity, at biscoitos's orchards, terceira island. **Proceedings of the 1st Congress of Fruit Growing and Viticulture, 2006.** Available at: <http;//www.azoresbioportal.angra.uac.pt/.../publications-congress-25- 33. Accessed on: 3 August 2012.

FIORANÇO, J. C. Brazilian apple: from import to self-sufficiency and export - technology as a determining factor. 2009. Available at: ftp//ftp.sp.gov.br/piea/publicações/E/2009/tec6-0309.pdf.

Accessed on: 3 August 2012.

FREE, J. Rosaceae: Malus, prunus and pyrus. In: **Insect Pollination of Crops.** 2nd Edition University of Wales, Cardiff, UK, Academic Press, 1993. Chap. 57, P.431 - 466

FREITAS, M. B. Bees as pollinating agents in food production and conservation of floral resources. Proceedings of the 43rd Annual Meeting of the SBZ-   JoãoPessoa-2006   .   Available at   :
<http;//www.files.cesaiifce.webnode.com.br/.../...Acesso on: 15. Dec. 2011.

IMPERATRIZ-FONSECA, V. L. Ecosystem services with an emphasis on pollinators and pollination. 2004. Available at: <http;// www.Ib.usp.br. Accessed on: 30 December 2012.

JAMES, R.; SINGER, T. Bees in nature and on the farm. In: **Bee Pollination in Agricultural Ecosystems.** Oxford University Press, 2008, Chap 1, p.3-9.

JOSHI, N. C.; JOSHI, P. C. Foraging behaviour of Apis *Spp.* on apple flowers in a subtropical environment. New York Science Journal, n.3, v.3, p.71-76. 2010.

KEOGH, R. et al. Apples. Australian Government, Rural Industries Research and Development Corporation Publication No. 10/109, Horticulture Australia Limited, 2008.

KREMEN, C. et al. The area requirements of an ecosystem Service: Crop Pollination by Native Bee Communities in California. Blackwell Publishing Ltd/cNRS, **Ecology Letter,** n.7, p.1109-1119, 2004.

KREMEN, C. Crop pollination services from wild bees In: **Bee Pollination in Agricultural Ecosystems.** Oxford University Press, 2008, Chap 2, p.10-26.

LOPES, P. R.; OLIVEIRA, I. V. Temperate climate fruit production in the Brazilian semi-arid region. 2010. Available at   :<http;// www. alice cnptia.embrapa.br/handle/doc/859903. Accessed on: 3 Aug.2012.

MAUÉS, M. M. Reproductive phenology and pollination of the brazil nut tree (Bertholletia excelsa Humb. & Bonpl. Lecythidaceae) in Eastem Amazónia, in: Kevan P & Imperatriz Fonseca VL (eds) - Pollinating Bees - The Conservation Link Between Agriculture and Nature - Ministry of Environment / Brasília, p.245-254. 2002.

MELLO, L. R. Brazilian apple production and market-panorama 2005. Available at: <http: www.infoteca.cnptia.Embrapa.br/handle/doc/541865. Accessed on: 3 Aug.2012.

OLIVEIRA, R.; SCHLINDWEIN, C. Limited fruit production in *Hancornia speciosa* (Apocynaceae) and pollination by noctumal and diumal insects. Biotropica (Lawrence, KS), Lawrence, v.37, n.3, p.381-388, 2005.

PETRI, J. et al. Advances in apple growing in Brazil. **Rev. Bras. Frutic.** - São Paulo - Special Volume, E. 048 - 056 - Oct 2011. Available at: <http;//www.scielo.br/pdf/rbf/v33nspel/ao7v33nspel.pdf. Accessed on: 2 August 2012.

RICHARDS, A. J. Plant breeding systems. Chapman and Hall, 2nd edition. P.529.1997.

STERN, R. et al. The appropriate management of honey bee colonies for pollination of rosaceae fruit trees in warm climates. Middle East and Russian **Journal of Plant Science and Biotechnology,** 2007.

TODD, F.; MCGREGOR, S. The Use of honey bee in production of crops. Annu. Rev. Entomol. Research Division, Agricultural Research Service, United States Department of Agriculture, v. 5, p.256-278. 1960.

VENTURIERI, G. Floral biology and management of stingless bees to pollinate assai palm (Euterpe oleracea Mart., Arecaceae) in eastem amazon. In: Pollinators Management in Brazil. Ministry of the Environment, Brasilia, 2008.

VIANA, B. F.; SILVA, F. O.; ALMEIDA, A. M. Yellow maracuja pollination in the semi-arid region of Bahia. In: Project management plans: sustainable use and restoration of native pollinator diversity in agriculture and related ecosystems. Ministry of the Environment (MMA)/ PROBIO (unpublished).

CHAPTER 1

Floral biology and pollination deficit in apple trees *(Malus domestica* **Borkh) in Chapada Diamantina, Bahia**

Abstract: Apple trees *(Malus domestica)* of the Eva variety grown in the northeast are economically promising and supply the domestic market, but productivity is lower than in other regions of Brazil. In order to understand the requirements for pollination and to assess the pollination deficit of this variety, the functional morphology and floral density, the density and diversity of floral visitors and the pollination deficit were described, comparing the number of seeds in the fruit from natural pollination and manual cross-pollination. The experimental units were established in commercial orchards of the Eva variety (commercial) in Cascavel (13°24'50, 7"S and 41°17'7.4"O), district of Ibicoara, Chapada Diamantina, BA. The flowers are gathered in inflorescences, are dialipetalous, pink, conspicuous, plate-shaped, bisexual, nectariferous and have daytime anthesis. The nectar produced in the centre of the flower is accessible to bees visiting the flowers. Flower density was higher in the Princesa cultivar compared to Eva (ANOVA a = 0.05; $F_{I,I24}$ = 21.63; p < 0.001). In 2010, 16 species from four insect orders were recorded (H' = 0.8316 species) and the density of visits did not differ between the Eva and Princesa cultivars (a = 0.05; KW = 1.10; p = 0.33). *Apis mellifera* L. was the most abundant species and an efficient pollinator, while the other species were scarce and not very effective at pollination. However, the efficiency of A. *mellifera in pollination* is greater when it forages for pollen, and less likely when it collects nectar, and the distances between the location of the hives and the apple trees influences the rate of visitation to the flowers. The pollination deficit was evidenced by the low number of seeds per fruit (average = 4 seeds/fruit) resulting from natural pollination (a = 0.05; $F_{I,32}$ = 33.09 ; p < 0.001). Densification with 5 hives/ha does not alleviate the pollination deficit, which is responsible for the low productivity in the region.

Keywords: agroecosystem, cerrado, Eva cultivar, seed formation, *Apis mellifera.*

INTRODUCTION

The apple tree *(Malus domestica)* is dependent on pollinators for the formation of seeds and fruit (Hegedus, 2006; Klein *et al.,* 2007). Bees are the main pollinators, and the reduction in productivity in apple orchards is related to the reduction in the number and diversity of these insects (Freitas *et aL,* 1995; Kevan, 1997; Kevan and Phillips, 2001), due to habitat loss promoted by intensive agriculture, pathogens, parasites, invasive species and the use of pesticides (Biesmeijer *et al.,* 2006). This is why the management of *Apis mellifera* hives in the orchard is an important factor for productivity and improving fruit quality (Freitas *et al.,* 1995; Kevan, 1997).

Pollination can fail at various points during the process of distribution, transport and deposition of pollen on the stigma (Thompson and Goodell, 2001), related to the mechanisms of pre-dispersal, dispersal and post-dispersal of pollen grains (Neiland and Wilcock, 2002). Due to self-incompatibility, the formation of fruit from natural pollination in apple orchards requires the intercropping of at least two genetically different varieties (Denardi and Camilo, 1986; Hegedus, 2006), so that one of them provides compatible pollen for the commercial variety. Factors such as the insufficient number or distance from the pollinating variety and the reduced number of visits to the flowers by pollinators can result in insufficient pollen being deposited on receptive stigmas of the producing variety (Kevan, 1975; Matsumoto *et al.,* 2008; Vaissière *et al.,* 2011).

The reduction in productivity of apple orchards due to inadequate pollen deposition, which characterises a pollination deficit, can be measured qualitatively (in relation to fruit and seed characteristics) or quantitatively (in relation to production). Thus, the existence of pollination failure can be demonstrated by pollination experiments in which seed and fruit production is evaluated in relation to the addition of pollen (hand-pollination experiments *versus* natural pollination) or the pollinator (densification of orchards with beehives) (Thompson and Goodel, 2001).

In Brazil, apple production began in the 1970s and is concentrated on two cultivars, 'Gala' and 'Fuji', which account for around 90% of the planted area (Soster and Latorre, 2007; Petri *et al.,* 2011). Brazilian apple production accounts for 1.5% of world production (Petri *et al.,* 2011) and the projected harvest for 2011/2012 is 1,366,000 tonnes (AGE, 2012). The huge boost in production, estimated at 6000% over the last two decades, has been favoured by technological innovations and improved cultivation practices, as well as the expansion of apple-producing areas into non-traditional areas towards warmer regions of the country (Petri *et al.,* 2011). In the Vale das Vertentes in Minas Gerais and the Chapada Diamantina in Bahia, the Eva (commercial) and Princesa (pollen donor) varieties are planted, selected for their suitability to local climatic conditions, requiring only a few hours of cold to break dormancy and grow the vegetative branches (Petri *et al.,* 2011).

In the Chapada Diamantina, apple production is 315 tonnes, with productivity of just 7.5 tonnes/ha, even with the practice of densification with beehives, suggesting that there is a pollination deficit. This productivity is lower than that observed in other regions of the country for the same variety, where it is currently close to 40 tonnes/ha, with some orchards producing over 50 tonnes/ha (Petri *et al.,* 2011).

Despite the geographical expansion and intensification of planting, information on the morphology and biology of the flowers of the Eva and Princesa varieties is non-existent, making it difficult to understand their influence on pollinator behaviour and, consequently, on the pollination deficit. This study aims to fill this gap, emphasising the description of aspects of floral biology, flower density and the pollination system. The number of seeds produced per fruit in the manual cross-pollination and natural pollination treatments will be used to assess the qualitative pollination deficit in an orchard with 5 *Apis mellifera hives/ha*. The number of species and the density of floral visitors in the orchard are correlated with seed production.

MATERIAL AND METHODS

Study area

This study was carried out in apple orchards of the Eva (commercial) and Princesa (pollinator) varieties belonging to the Bagisa S/A Agropecuária e Comércio company, part of the Ibicoara-Mucugê Agropolo, located in Cascavel (13°24'50, 7"S and 41°17'7.4"O), a district of Ibicoara in the Chapada Diamantina, BA (figure 1).

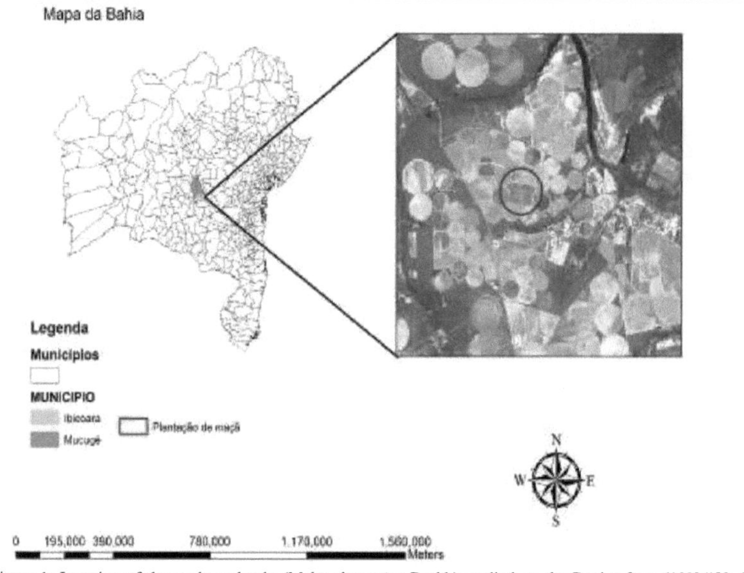

Figure 1: Location of the apple orchards *(Malus domestica* Borkh) studied on the Bagisa farm (13°24'50, 7"S and 41°17'7.4"O). Supervised high-resolution satellite image (SPOT), taken in September 2008, of the Cascavel district, Ibicoara, Bahia, Brazil.

The 19 enterprises that make up the Ibicoara-Mucugê Agropólo occupy 200,000 hectares, responsible for producing more than 50% of the potatoes consumed in the northeast, as well as tomatoes, cabbage, onions, garlic, pumpkins, peppers, beans, apples, plums and other crops. In the area occupied by the Agropólo, conventional cultivation is predominantly practised, with a centre pivot irrigation system.

The predominant physiognomy in the region is of the savannah type, with an altitude of approximately 1100 metres and an average annual temperature of 21°C, with average highs of 26°C and average lows of 16°C. The rainy season runs from November to March, with annual rainfall of 757 mm (Source: monitoring data provided by Fazenda Bagisa).

The apple orchard

The 43 hectares of Fazenda Bagisa's commercial orchard are divided into 36 plots, organised into three blocks 10 metres apart (Figure 2). The blocks identify the 2006 planting (500 05) and the 2007 planting (500 06), each with 18 plots of 1.2 hectares each. The plot is made up of 14 rows, with a spacing of 4 metres between plants, with 5 Evas to one Princesa (5:1 ratio), the latter being the pollen donor, representing around 10-12% of the orchard.

The orchard is managed using the conventional system, with applications of soil improvers, fertilisers and pesticides. A mixture of 0.75 TA 35 + zinc + 11.25 copper sulphate is used to remove the mature foliage after harvest, preventing the buds from going into endo-dormancy, and the products 12 dormex (hydrogen cyanamide), 0.75 pyrimex, 60 Agrex oil are used to induce flowering. These are mixed to form a 15,000 litre mixture which is then applied to an area of 1.2 hectares.

The investigations into floral and reproductive biology and the behaviour of floral visitors were carried out in a 2-hectare orchard different from the one described above. The tests to assess the pollination deficit, quantify flower density and the density and diversity of visitors were carried out in the eight sampling units drawn from the 36 plots in the 43-hectare commercial apple orchard. Each sampling unit is 25 m wide x 50 m long, preferably located in the centre of each plot, with a minimum distance of 10 m from the edge (Figure 2).

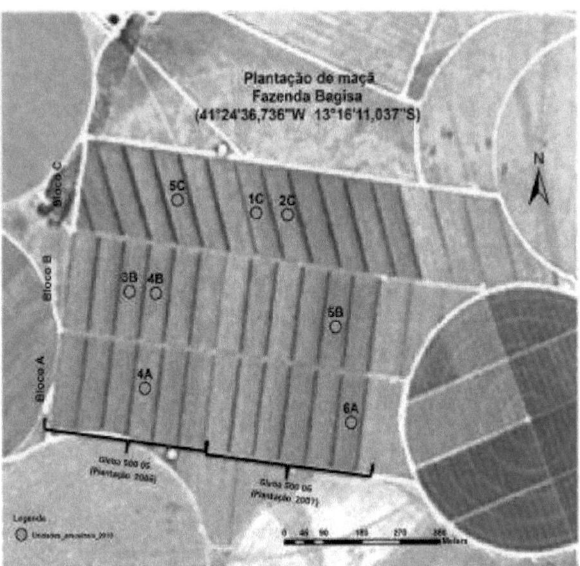

Figure 2: Top overview of the apple orchards in the study area (43 ha) with the subdivisions into two Glebas (500 05 and 500 06) at Fazenda Bagisa, Cascavel district, Ibicoara, Bahia, Brazil. The green circles (1C, 2C, 5B, 6A, 5C, 4B, 3B) indicate the experimental plots used to study the pollination deficit in 2010. Supervised high-resolution satellite image (SPOT - pixels 5 m on a side) taken in September 2008.

Apple varieties

The Eva variety is the product of a cross between the Gala and Ana varieties, made in 1979 by the Agronomic Institute of Paraná, with low chilling requirements - 300 to 350 hours - and early fruit ripening compared to the Gala variety (IAPAR, 2009). The Eva variety has medium to high vigour and is highly productive, with striped red fruit on a whitish-yellow background, with a good appearance. They have a sweet, sub-acid flavour, a satisfactory shelf life of 7 to 10 days, a conical to rounded shape and a slightly larger size than the Gala variety. The Princesa cultivar, developed by Epagri in Caçador-SC, is used exclusively as a pollen donor because it has a low fruit yield but a high volume of flowers with prolonged flowering, producing a large quantity of pollen.

Biology and floral morphology

The Eva and Princesa varieties were characterised in terms of floral morphology using samples of flowers kept in 70% alcohol under a stereoscopic microscope in the laboratory. Additional information was obtained from photographic records of the phenophases in the field. All the procedures below were carried out according to Dafni et al. (2005). The main morphological and physiological changes during the phenophases were described based on the observation of five flowers, distributed on five plants of each cultivar. Stigma receptivity was tested with hydrogen peroxide (n=5 flowers per phenophase), considering the formation of bubbles on the stigma as an indication of receptivity. To test pollen viability (n=10 flowers per phenophase), neutral red (1%) was used to stain the protoplasm of the pollen grains, which is indicative of viability.

The flower buds (n=25 flowers for each interval) were bagged the day before the measurements to prevent visits. Nectar samples were taken every four hours: at 9am, 1pm and 5pm on (day 1 of anthesis); at 1pm and 5pm (day 2° of anthesis) and at 5pm (day 3° of anthesis). The sampling time was chosen because of the greater availability of open flowers at this time, despite the fact that opening begins at 7.30am. The sugar concentration in the nectar was measured using a refractometer (from 0 to 90%). To calculate the mg sugar, the average amount of nectar (volume in pL) was multiplied by the sugar concentration present in one pL of nectar, according to the conversion table in Dafni et *al.* (2005). The nectar was extracted successively from flower samples using 5.0 pL microcapillaries, assessing the volume, production pattern and the effect of removal by visitors on one-day, two-day and three-day flowers (n = 25 flowers for each treatment).

Pollination experiments

In 2008, Freire (2009) carried out pollination experiments on the Eva and Princesa varieties to test apomixis (TI - removal of anthers on flowers kept bagged), spontaneous self-pollination (T2 - flowers bagged without manipulation), manual self-pollination (T3 - pollen transferred to the stigma of the same flower), manual cross-pollination (T4 - pollen from Princesa flowers transferred to the stigma of Eva flowers), natural pollination (control) (T5 - flowers open to visitors and not handled), geitonogamy (T6 - transfer of pollen between flowers on the same plant, and

in different individuals of the Eva variety). The centre bud on each inflorescence ("king blossom"), which produces the best quality fruit and remains after thinning, was used for the experiments (West, 1999).

For each treatment, 80 flower buds were selected, one bud per inflorescence, distributed among the 20 selected individuals (4 buds per plant) arbitrarily during peak flowering. The treatments (Tl, T2 and T3) were carried out on both varieties and the treatments (T4, T5 and T6) were only carried out on the Eva variety. With the exception of the control (T5), whose buds were only marked, in the other experiments the flower buds were kept bagged with nylon mesh bags until the manipulations were carried out, i.e. until anthesis.

In order to assess the effect of distance from the hives on the number of fruits and seeds on the apple trees, five trees of the Eva variety were drawn (sample points), distributed between the minimum distance from the hives of 4m and the maximum of 192m, according to the management used by the producer. The A. *mellifera* hives were previously installed in the orchard in four groups of 7 hives, randomly distributed within the plantation. After 20 days, the developing fruit was collected to count the number of fruits and the number of seeds per fruit.

Evaluation of pollination deficit

In 2010, the methodology described in the "Protocol to Detect and Assess Pollination Deficit in Crops - FAO/IFAD" (Vaissière *et al.*, 2010, 2011) was adopted for studies in apple orchards to assess pollination deficit, flower density and the density and diversity of floral visitors. This protocol also guided the delimitation of the number and location of plants sampled in the experimental units (Figure 2). This year, the grower started using 5 hives/ha arranged in groups between rows 13 and 14, i.e. at the edge of each plot.

In each of the eight experimental units selected, 16 plants were chosen and distributed into 4 plots, each consisting of 2 Eva and 2 Princesas, totalling eight plants of each cultivar (Figure 3). On each individual plant, 2 buds (the central flower of the inflorescence) were selected in similar positions but in different bouquets, which were marked and bagged until anthesis, one bud for the natural pollination test and the other for manual cross-pollination. For cross-pollination, the dehiscent anthers of Princesa flowers were rubbed lightly against the receptive stigmas of Eva flowers, observing the deposition of pollen grains on the stigmas using a magnifying glass (lOx). Naturally pollinated flowers were re-bagged after 24 hours and cross-pollinated flowers were re-bagged immediately after pollen transfer. Both flowers remained bagged for 90 days, when they were collected to count the number of seeds.

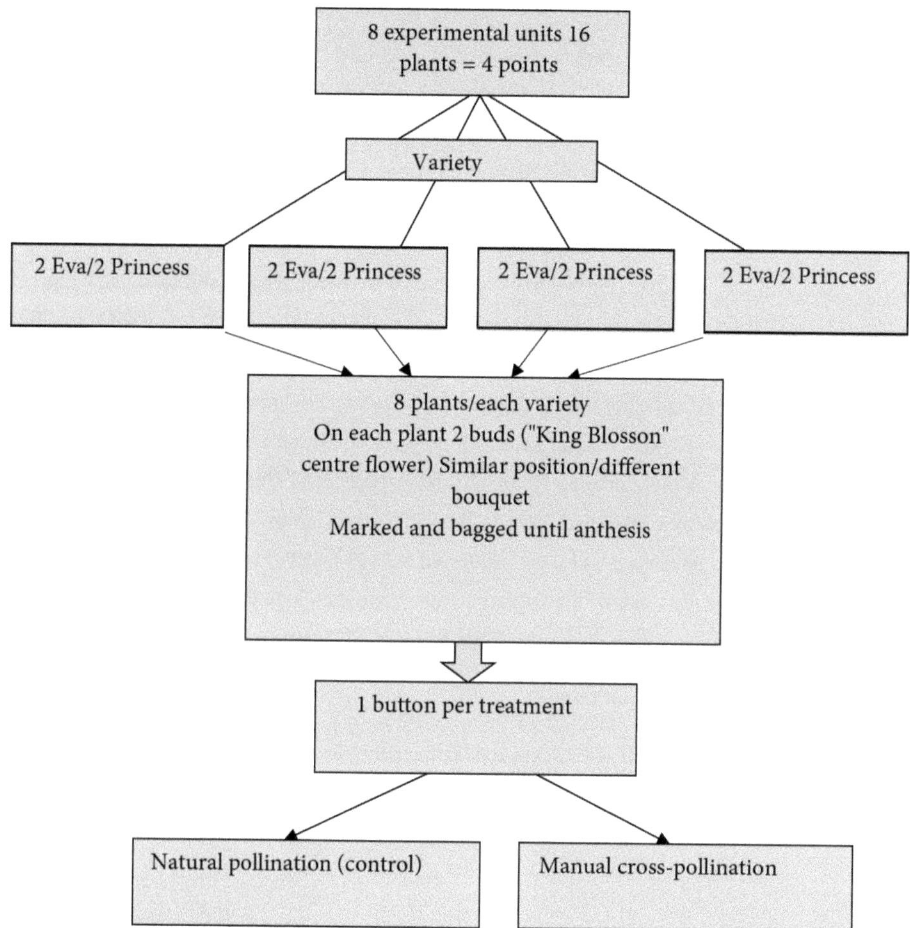

Figure 3: Schematic representation of the procedure carried out to assess the apple tree pollination deficit in 2010.

Assessment of flowering density

Recording was done once in each of the eight experimental units on the 16 selected plants, distributed in 4 plots marked out in rows in the centre of the experimental area. Each plot consisted of two Eva and two Princesa plants, totalling eight plants of each cultivar. The flowers on each individual were quantified on 2 opposite branches, representative of the plant, with a base width of around 4-5cm. Flower production was estimated by multiplying the average number of flowers by the total number of branches on the plant (Vaissière *et al.*, 2010).

Assessing the density of floral visitors

Visitor density was quantified in the same plots used to assess flower density. The records were made by a collector travelling through the rows in the morning (starting at 9am) and afternoon (starting at 2pm) for two consecutive days, during the orchard's most intense flowering phase according to the protocol. Two counters were used to record the number of pollinators seen on each floral unit. Fifty flowers were used for each individual, this number being defined according to the availability of flowers on the plants. This assessment was carried out in good weather conditions to allow the bees to forage (Vaissière *et al.,* 2010).

Assessing the diversity of floral visitors

In each of the eight experimental units, 12 plants were selected and distributed into 6 plots (5 plots made up of two Eva cultivar plants and 1 plot made up of two Princesa cultivar plants), totalling ten Eva cultivar plants and two Princesa cultivar plants. The plots were located on the edge of the experimental area, so the collector travelled around the plots and around the experimental area in the morning (starting at 10am and in the afternoon at 2pm). Collections were made with an entomological net for 5 minutes in each plot, totalling 30 minutes per experimental area according to the protocol. The collection intervals were from 9am to 4pm. The captured specimens were placed in a mortifying chamber containing ethyl acetate and then transferred to vials labelled with the date of capture, the experimental area and the name of the collector. They were then kept under refrigeration (-20°C), subjected to the mounting steps and identified (Vaissière *et al.,* 2010).

Characterisation of the surroundings

The proportion of habitat and the non-spatial diversity (Shannon index) of habitat types around the apple orchards were calculated using the Patch Grid application together with ArcGIS 9.3, based on a circular area with a radius of 2km, starting from the centre of the commercial orchard at Fazenda Bagisa.

Statistical analyses

To analyse the difference in the total number of seeds produced per fruit between the free pollination and hand pollination treatments, we used the ANO VA test. The same was done to test the difference between the floral densities of the two varieties. Due to the heteroscedasticity of the data, i.e. the strong dispersion of the data from the density of floral visitors, it was submitted to the non-parametric Kruskal-Wallis test to test the null hypothesis of no difference in the mean number of floral visitors between the two varieties. All the analyses were carried out using *SPSS 19.0 for Windows* software.

To find out if there was a difference in the nectar production pattern of the cultivars in the three phases during floral longevity, the chi-squared test was used. All these tests were carried out using the GraphPad InStat programme.

RESULTS

The main morphological and physiological characteristics of the reproductive units in *Malus domestica* Borkh, in the Eva and Princesa varieties, are summarised in Table 1. The flowers are gathered in umbel-type inflorescences, monoclinous and showy, pentamerous, dialipetalous, green dialysepalous, actinomorphic; gynoecium, with an infertile ovary, five stigmas and five carpels each with two ovules; androecium with stamens of different sizes, twelve to fifteen yellow anthers with longitudinal dehiscence.

In the Eva variety, the petals are pinkish-white, with fused stipes at the base and longer than the stamens. Due to the longer floral petiole, the flowers in the bouquet are comparatively further apart than those in the Princess variety. In the Princesa cultivar, the flowers that form the bouquet are tightly joined; the sepals are hairy and the petals are bright pink when in bud and pinkish-white after the petals separate. The stipes are smaller than the stamens, which are fused at the base and have trichomes from the base to the apical third of the stipe (Figure 3).

Table 1: Description of aspects of the morphology and floral biology of the Eva and Princesa varieties in Chapada Diamantina, BA.

Floral characteristic	Parameters	variety Eva	variety Princess	Sample size
MORPHOLOGY				
Inflorescence	Length (cm)	5a6	4a5	10
Bouquet	Number of flowers	5a6	3a6	10
Stamens	Number	19 a 25	20 a 25	10
Stigma	Number	5	5	10
Peciolo	Length (cm)	1,5	0,5	10
PHYSIOLOGY				
Amount of Nectar	Average volume over the life of the flower (pl)	1.35 (day 1), 1.56 (day 2º) and 0.36 (day 3)	1.16 (day 1), 1.29 (day 2) and 0.29 (day 3)º	25
Sugar concentration in nectar	Mg of sugar during the life of the flower	0.6 (1st day), 0.63 (2nd° day) and 0.09 (3rd day)*	0.46 (day 1), 0.51 (day 2º) and 0.08 (day 3º)*	25
Nectar production	Post-collection replacement	It doesn't happen	It doesn't happen	25
Inflorescence longevity	Days	4 (max)	4 (max)	10
Flower longevity	Days	3	3	10
Pollen viability	Time in relation to floral longevity	The whole period	The whole period	10
Stigmatic receptivity	Time in relation to floral longevity	The whole period	The whole period	5
Anthesis	Timetable	7.30am to 5.30pm (mostly between 10am and 2pm)	7.30am to 5.30pm (mostly between 10am and 2pm)	10

* cannot be statistically tested due to insufficient number of samples

The chronological sequence of the main morphological and physiological changes in individual apple blossoms during longevity is summarised in Table 2. In each inflorescence, the flower in the centre of the bouquet opens first and then the other flowers open sequentially, in a non-chronological manner. Anthesis takes place throughout the day from 7am to 5.30pm and is most intense between 1pm and 2pm. Daily anthesis intensifies at times of higher temperature, favouring the visitation of bees, which forage more intensively at warmer times.

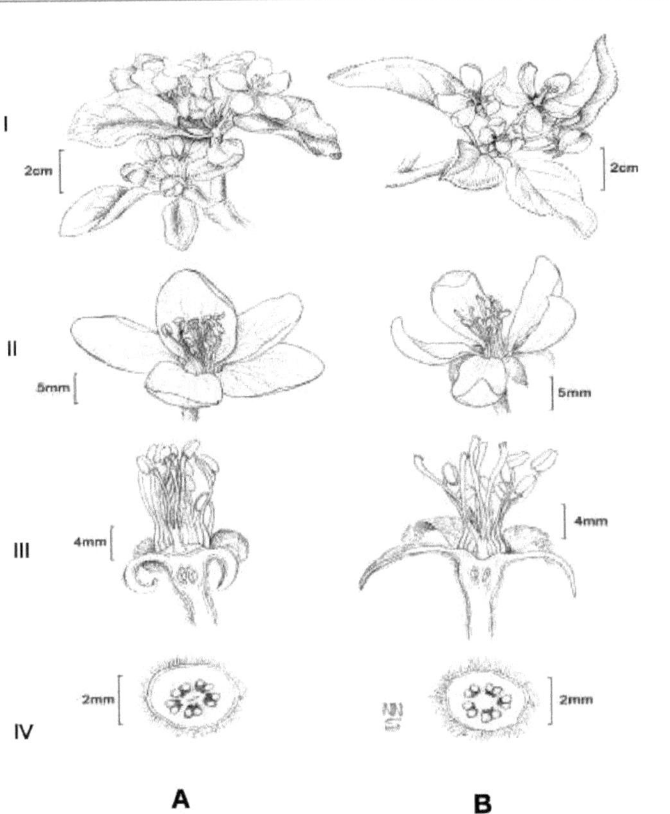

Figura 4: Representation of the two varieties of apple trees (A - Princesa and B -Eva) from the point of view of biology (Author of the board: Natanael Nascimento) and the stages referring to the phenophases: I- flowering or floral anthesis (opening of the flowers of the inflorescence, slightly separated sepals make the petals visible); II- individual flowers in each inflorescence; III-flowers without petals can be observed the arrangement of the stamens and pistils; IV-transverse cut at the height of the ovary where it is possible to observe the presence of locules.

In both varieties, the stigmas are receptive from the moment the flowers begin anthesis and remain so until senescence. The pollen grains are viable from the moment the flower opens until senescence, and in the same flower, the anthers of the longer stamens become dehiscent before the anthers of the shorter stamens.

Table 2. Sequence of morphological and functional changes identified in *M. domestica* flowers in Chapada Diamantina, BA during 2008.

Phenophase	Duration (hours)	Events
1	1h<	Flowers in pre-anthesis (beginning to open), intense pink petals, light yellow anthers, receptive stigma, greenish in colour;
2	1-3h	One petal opens and the others open slowly and sequentially; few dehiscent anthers with viable pollen;
3	4-8h	Pale pink petals perpendicular to the ovary, rosulate filaments and dehiscent anthers with viable pollen;
4	24h	Flowers still young, when not pollinated, pinkish-white petals, stigma still receptive, with some dark and aged anthers, but with other dehiscent anthers and viable pollen;
5	30h	More anthers in dehiscence, petals still open and pinkish-white in colour;
6	48h	Predominantly pale-pink petals; dark, aged anthers and still receptive stigma;
7	64h	Flowers with dark anthers and filaments that turn from pink to pale in colour
8	72h	The petals fall and, when pollinated, the sepals close to form the fruit; if not pollinated, the sepals remain open until the receptacle falls;

The Eva variety produced more nectar than the Princesa variety, although not significantly different (p=0.7507) (Table 3), and the same trend was observed for sugar concentration. However, the amount of nectar present in the flowers after the first removal was very small (Table 3).

Table 3. Average nectar volume in pL in Eva and Princesa cultivars measured in 3 groups (G1 = I° day of anthesis, G2 = 2nd day of anthesis and G3 = 3rd day of anthesis) with treatments (C1 = nectar collection from the same individuals at times: 9:00,13:00 and 17:00; C2 = nectar collection on the same individuals at times: 13:00 and 17:00; C3 = nectar collection on the same individuals at time: 17:00), in Chapada Diamantina, BA during 2008,

Cultivar	time	Flowers on the 1st day of anthesis			Flowers on the 2nd day of anthesis			Flowers at 3° day of anthesis			Number of flowers
		C1	C2	C3	C1	C2	C3	C1	C2	C3	
Eva	09:00	1,35	-	-	1,56	-	-	0,36	-	-	25
	13:00	0	1,49	-	0	1,56	-	0	0,14	-	25
	17:00	0	0	1,31	0	0	1,32	0	0	0,11	25
Princess	09:00	1,16	-	-	1,29	-	-	0,29	-	-	25
	13:00	0	1,05	-	0	1,29	-	0	0,17	-	25
	17:00	0	0	0,91	0	0	0,80	0	0	0,11	25

The pollination tests carried out in 2008 did not produce any fruit in the apomixis and spontaneous self-pollination treatments. The comparison between the treatments with naturally pollinated flowers (open to visitation) and manual cross-pollination (table 4) revealed significant differences in both the number of fruits produced (unpaired t-test: p < 0.0001) and the average number of seeds per fruit (unpaired t-test: p = 0.0131).

In the study area, there was no significant relationship between the distance from the hives and the number of seeds in the fruit (p = 0.1244; (r) = -0.7741), but the visitation rate of *Apis mellifera* decreased as the distance from the hives increased (p = 0.0152; (r) = -0.9453). Despite the lack of statistical significance, the data shows that the number of seeds in the fruit is influenced by the distance between the hives and the orchard plants.The natural pollination and manual cross-pollination tests in 2010 reaffirmed the 2008 results, recording a significant difference in the number of seeds produced per fruit between the treatments ($\alpha = 0.05$; $F_{I,3}$ 2 = 33.09 ; $p < 0.001$) in the apple orchards (Figure 4). The points on the graph indicate the values found on individual plants and are reduced due to the high number of abortions (only 15 values could be counted). Due to the high number of abortions in the Princesa variety, which occurred during the period of the experiment, the sample size was not satisfactory for analysing the results for this variety.

Table 4. Total number of fruits produced and average number of seeds per fruit formed in the manual cross-pollination and natural pollination tests, in relation to the distance from the hive clusters, in a *Malus domestica* Borkh plantation in the Chapada Diamantina region, BA during 2008.

Distance from the hive cluster (m)	Manual cross-pollination		Natural pollination	
	No. of seeds per fruit *	Fruits *	No. of seeds per fruit *	Fruit*
4	5,2	59	5,1	36
56	5,1	63	3,6	41
114	4,6	65	3,6	32
134	5,5	62	3,6	34
192	4,6	58	3,5	36

*Significant differences: Fruits (p < 0.0001) and No. of seeds per fruit (p = 0.0131).

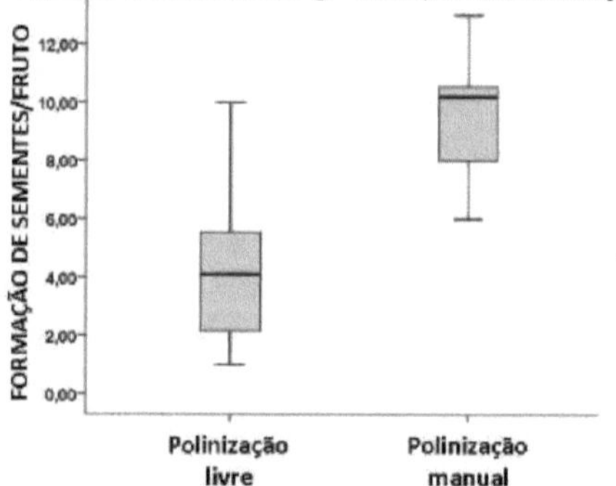

Total number of seeds produced per fruit in *Malus domestica* Borkh in the natural pollination and manual cross-pollination treatments for the Eva cultivar, carried out in 2010 in Chapada Diamantina, BA.

The density of visitors to the flowers of the Eva variety averaged 6.98 individuals/50 flowers (SD = ±2.55) and the Princesa variety averaged 5.17 individuals/50 flowers (SD = ±4.59). There was no significant difference between the varieties in terms of average visitor density ($\alpha = 0.05$; KW = 1.10; p = 0.33).

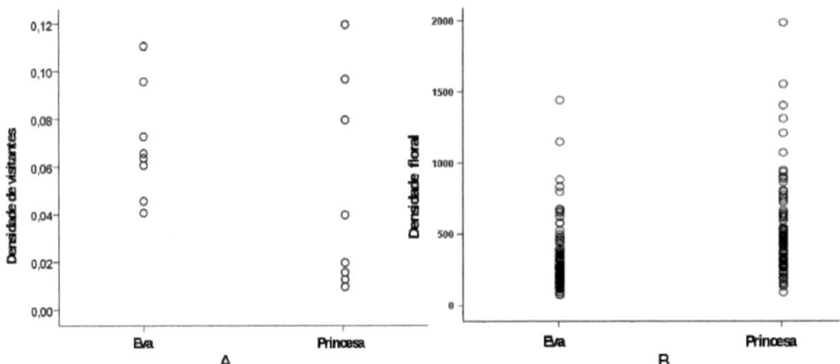

Figure 6: Density of floral visitors (A) and flower density (B) in the Eva and Princesa varieties in apple orchards, recorded in 2010 in Chapada Diamantina, BA.

Table 5: Diversity of floral visitors sampled with entomological nets in apple orchards in Chapada Diamantina, BA during 2010.

Order	Species	Number of individuals collected
Hymenoptera	*Apis mellifera* Linnaeus, 1758	256
	Trigona spinipes Fabricius, 1973	2
	Xylocopa grisescens Lepeletier1841	1
	Brachygastra lecheguana Latreille, 1824	1
	Vespidae sp.2	3
	Vespidae sp.4	1
Diptera	Anthomyiidae sp.1	4
	Bibionidae sp.1	1
	Chrysomelidae sp.1	3
	Chrysomelidae sp.2	3
	Muscidae sp.1	22
	Syrphidae sp.3	1
	Syrphidae sp.6	5
	Syrphidae sp.9	3
Lepidoptera	Lepidoptera sp	3
Hemiptera	Hemiptera sp.1	1

On the other hand, flower density was significantly higher in Princesa than in Eva (ANOVA for one factor: $\alpha = 0.05$; $F_{1,124} = 21.63$; $p < 0.001$), where the Princesa variety had a higher flower density than the Eva variety (Figure 6). The flower:fruit ratio for the Eva variety (per individual) is 154:80 fruit and for the Princesa variety this ratio is 242:50 fruit.

During the 2010 flowering season, 16 species representing four orders of insects visited apple blossoms in the Chapada Diamantina (Table 5), characterising low diversity (H' = 0.8316 species). Bees were represented by three species, with *A.mellifera* being the most abundant, while the other bee species were scarce in the orchards.

In the area covered by this study, the landscape around the orchard is not conducive to colonisation and the provision of trophic resources for pollinating bees. Within a radius of 2 km, starting from the centre of the orchard, the agricultural matrix represents 76.7% of the surroundings, so the proportion of habitat represents only 23.3%. The Shannon diversity index is equal to 0.4, which indicates a low level of heterogeneity in the landscape.

Aspects of the foraging behaviour of *Apis mellifera* bees, such as their pronounced constancy in the flower, carrying significantly heavier pollen loads and in greater quantities (Mattu, 2012), indicate their effectiveness in transferring pollen between varieties in orchards. The other nectariferous insects collected from flowers are not effective pollinators of apple trees due to their low efficiency. Based on observations in the field, we can say that they make illegitimate visits and do not pollinate, with the exception of *Trigona spinipes*.

The interaction between flower morphology and nectar or pollen collection behaviour determines the contact between the pollinator's body and the stigma. During the sampling period, *A. mellifera* collected comparatively more nectar than pollen (p = 0.0019), resulting in a lower frequency of visits with contact between the visitor's body and the anthers and stigma (p = 0.0124). To collect nectar, the bees land on the petals, move towards the centre of the flower and insert the glossa into the receptacle, where the nectar is produced and accumulates. The bee's head contacts the base of the stamens and stipes and, rarely, the stigmas and anthers.

On the other hand, during active pollen collection the bees land and move around on the anthers and consequently contact the ventral region of the body with the stigmas. In addition, the time spent visiting a flower is longer when bees are collecting pollen, which also increases the chances of efficient pollination. In this way, pollination is favoured by foraging for pollen and, less frequently, by foraging for nectar.

DISCUSSION

The flowers of the Eva and Princesa varieties, despite some peculiarities, have the typical morphology of the Maloideae subfamily (Rohrer *et al.,* 1994). However, many aspects that influence the attractiveness of visitors and pollination in these varieties are still unknown, compared to other traditionally marketed varieties (Sheffield *et al.,* 2005). The great longevity of the flowers and inflorescences contributes positively to increasing the chances of effective pollination during the short flowering period of the plants in the orchards.

The fact that the flowers remain open at night and the drastic reduction in nectar volume in senescent flowers with the exclusion of visitors (3 days) suggests the existence of resorption. However, this aspect was not tested in this study and there are no previous records for apple trees, although there are reports for other plant species where nectar reabsorption at night and daily production is interpreted as an energy-saving strategy (Cruden *et al.,* 1983). The lack of nectar replenishment in Eva and Princesa differentiates them from other apple varieties studied where nectar replenishment occurs (Freitas, 1995). According to this author, nectar replenishment in these varieties increases their attractiveness to pollinators and, consequently, their chances of being pollinated. It is possible that in the Eva variety, the reduction in nectar production in 3-day flowers may be compensated for by the morphological similarity, prolonged receptivity of the stigma and asynchrony in flowering, which keeps flowers in the inflorescences at different stages of anthesis.

The orchard pollination deficit detected in 2008 and 2010 in the Eva variety may be the result of deficiencies in the management of the hives and the pollinator variety, especially with regard to distance from the hives and the flowering of the pollinator, factors pointed out by other authors as crucial for efficient pollination in apple orchards (Kevan, 1975; Matsumoto *et al.,* 2009) and therefore for productivity in apple orchards.

In the orchard studied, the pollinating variety is in the recommended proportion, based on studies of pollen dispersal in apple orchards carried out by Kevan (1975) and the variety used (Princess) has suitable characteristics such as pollen compatibility and viability, high flower density, annual rather than biannual flowering, simultaneous or overlapping flowering with the producing variety, distinct fruit at the harvest stage and occupation of a small columnar space (Dennis, 2003).

However, the management practised by the grower to induce flowering in the area studied promotes early flowering of the donor variety, so that the peak flowering of the commercial variety occurs during the decline in flowering of the donor variety, with negative effects on pollen transfer, foraging and pollinator behaviour, maintaining low seed production. The early flowering of the pollinated variety is expected and, in order to control it, the producer first applies the inducing hormone to the producer variety and, days later, to the pollinating variety. This flowering behaviour in response to the induction may be due to the water stress to which the plants are subjected during prolonged periods of drought in the region and the age of the plant, since this procedure ensures that the flowering of the varieties overlaps in the first few years. This difference in flowering response between varieties increases harvesting costs, as part of the fruit is still developing at harvest time, forcing the producer to carry out the harvest in at least two stages.

Another important factor is the spatial arrangement of the *Apis mellifera* hives used to pollinate the orchard adopted by the producer in 2008 and 2010. In 2008, the use of 1 hive/ha arranged randomly in the orchard proved to be inadequate because the plants furthest from the hives are progressively less visited, consistent with studies that have reported a linear reduction in the number of honey bees as they move away from the hives (Free, 1993; Paranhos *et al.,* 1997), suggesting that the champions select floral sources closer to the hives, due to the lower energy cost of foraging. The hive management practised by the producer in 2010, with a standardised density of 5 hives/ha arranged at the edge of each plot, did not alleviate the pollination deficit in the orchard.

Although densification and spatial arrangement favour the predominance of A. *mellifera* due to the increased population density of these bees (Kevan, 1997). However, surveys of floral visitors in apple orchards in other locations have recorded a greater diversity of bee species, even in orchards densely populated with *Apis mellifera* (Gardner and Asher, 2006).

The low diversity and density of visits to flowers by native bees in this area is influenced by the low representation of natural habitats in the surroundings, as the region is dominated by intensive agriculture and semi-natural habitats. In this way, the scenario in the region studied agrees with empirical studies that relate a reduction in the diversity of floral visitors to the distance between orchards and natural areas (Rickets *et al.,* 2008; Klein *et al.,* 2008; Watson *et al.,* 2011). In addition, the presence of concomitant flowering of ruderals and other crops that are more attractive to apple trees, such as plums *(Prunus* sp.), in the surrounding area, may attract floral visitors and reduce visitation to apple tree flowers at the study site.

In the varieties studied, the amount of sugar in the nectar of individual flowers was lower than that recorded in other varieties, estimated at between 0.60 and 1.39 mg on average (Free, 1993), which may have increased foraging for nectar. Thus, flower-visiting behaviour is determined by the type of resource collected, which directly influences pollination efficiency. In addition to *A. mellifera,* species from the genera *Andrena, Bombus, Halictus* and *Osmia* are mentioned in the literature as efficient apple pollinators (McGregor, 1976; Vicens & Bosch, 2000; Sheffield *et al.,* 2008).

Among the floral visitors recorded in orchards, only A. *mellifera* is more viable for pollination, as is the case for other cultivated varieties in different regions of the world (Kevan, 1997; Thompson and Goodell, 2001). In Brazil, the role of native species in pollinating the Eva apple tree is still unknown. Although *T. spinipes* bees are considered harmful to certain crops (Silva *et al.,* 1997), in the apple tree they not only did not damage the flowers, but also made legitimate visits, with a potential similar to that described for *A. mellifera in* foraging for pollen and nectar.

The lack of emphasis on the role of native pollinators and the maintenance of friendly habitats for native pollinators keeps producers dependent on this species (Tepedino *et al.,* 2007), which is considered an input in agricultural production. It should be borne in mind that A. *mellifera,* like native bee species, are susceptible to exposure to pesticides used to control apple tree pests (Kevan, 1975). In this sense, the adoption of organic farming practices could mitigate the effects of agrochemicals on these bees, favouring the maintenance of pollinating bees in the orchard (Holzschuh, 2008).

Ongoing studies have identified 7 species of native bees that are potential pollinators of this variety of apple tree in the Chapada Diamantina. In addition to the introduced species *Apis mellifera scutellata* Lepeletier, 1836, the most likely to be used to pollinate apple orchards are: *Melipona quadrifasciata anthidioides* Lepeletier, 1836, *Melipona quinquefasciata* Lepeletier, 1836, Geotrigona subterrânea (Friese, 1901), *Exomalopsis analis* Spinola, 1853, *Xylocopa cearensis* Ducke, 1910 (Apidae), *Dialictus* sp. and *Augochlora sp.* (Halictidae) (data provided by the Crop Pollinator Monitoring Network, 2012).

Among the species mentioned above, *Melipona quinquefasciata, Geotrigona subterrânea* and *Exomalopsis analis* nest in the soil and are difficult to adapt for breeding. The species *Melipona quadrifasciata anthidioides* and *Xylocopa cearensis* proved to be the most promising to be reared for the directed pollination of apple trees, and techniques for their management and rearing are being tested. The nests of *Dialictus* sp. and *Augochlora* sp. have not yet been found in the area, which makes it difficult to understand their potential as good pollinators to be reared for the directed pollination of apple trees (POLINFRUT, project in progress).

CONCLUSION

In the absence of native species visiting the apple tree in population densities sufficient to supply the pollination "services" in the area studied, A. *mellifera* plays an important role in the pollination of apple blossoms. However, the densification levels used are not sufficient to alleviate the pollination deficit, which results in low productivity in the region.

The data suggests that improved management of A. *mellifera* hives and greater control of flowering in the pollinator and producer varieties, as well as the adoption of practices that are friendly to native pollinators, can mitigate the negative effects imposed by isolation and loss of natural habitat on the diversity and density of pollinators in the orchard. Greater attention needs to be paid to the quality of the hives used for pollination, prioritising the use of strong hives with a greater number of individuals.

This study has promoted advances in hive management applied by the producer, with the aim of reducing the negative effects of the distance between hives on bee activity and increasing the efficiency of pollen transfer between varieties.

Eva and Princesa. The centralised arrangement of hives within the crops increases the efficiency of A. *mellifera* in apple pollination, but was not enough to reduce the pollination deficit, suggesting that the density used is still insufficient to guarantee adequate pollination of the orchard. The diversity of the native bee fauna could increase pollination levels and reduce the demand for managed hives.

ACKNOWLEDGEMENTS

The authors would like to thank CNPq and GEF/FAO/FUNBIO/UNEP for their financial support, which made this research possible. Dr Favízia Freitas de Oliveira, from the Zoology Museum at the Federal University of Bahia, for identifying the insects collected from the flowers. The Bagisa S/A Agropecuária e Comércio company for its logistical support and access to the orchards. Blandina F. Viana thanks the CNPq research productivity grant. Fabiana O. da Silva thanks CAPES for the PNPD scholarship; Antonio da Costa Diakos thanks FAO/FUNBIO for the master's scholarship.

REFERENCES

STRATEGIC MANAGEMENT OFFICE. Brazil agribusiness projections 2011/2012 to 2021/2022. Ministry of Agriculture, Livestock and Supply. Available at: www.agricultura.gov.br//arq_editor. Accessed on: 29 October 2012.

BIESMEIJER, J. C. et al. Parallel declines in pollinators and insect-pollinated plants in Britain and the Netherlands. **Science,** v. 313, n.5785, p.351-354, 2006.

CRUDEN, R.W.; HERMANN, S.M.; PETERSON, S. Patterns of nectar production and plant-pollinator coevolution. In: BENTLEY, B.; ELIAS, T., (Ed.). **The Biology of Nectaries.** New York: Columbia University Press, p. 80-125,1983.

DAFNI, A.; KEVAN, P. G.; HUSBAND, B. C. Practical pollination biology. Canada: Enviroquest, p.590. 2005.

DENARDI, F.; CAMILO, A. P. EPAGRI-404-Imperatriz - new cultivar of apple tree. **Anais do XIV Congresso Brasileiro de Fruticultura,** Curitiba, Brazil, p.270.1986.

DENNIS, JR. F. 2003. Flowering, pollination and fruit set and development. In: FERREE, D. C.; WARRINGTON, I. J. (Eds.), Apples, Botany, Production and Uses. CABI Publish-ing, Oxon, UK, pp. 153-166.

FREE, J. B. Insect pollination of crops. London: Academic Press, p.684.1993.

FREITAS, B. M. The pollination efficiency of foraging bees on apple *(Malus domestica* Borkh) and cashew *(Anacardium occidentale* L.). Cardiff: University of Wales College, p.228.1995.

GARDNER K. E.; J. S. ASCHER. Notes on the native bee pollinators in New York apple orchards. **Journal of the New York Entomological Society,** 114(1):86-91. 2006.

HEGEDUS, A. Review of the self-incompatibility in apple (Malus x domestica Borkh., syn.: Malus pumila MilL). **Int. J. Hortic. Sei.** 12 (2), p.31-36. 2006.

HOLZSCHUH, A.; DEWENTER, I. S.; TSCHARNTKE, T. Agricultural landscapes with organic crop support higher pollinator diversity. **Oikos.** 117: p.354-361, 2008.

AGRONOMIC INSTITUTE OF PARANÁ. Available at: www.iapar.br > Accessed on: 20 January 2009.

KEVAN, P. G. AND T. P. PHILLIPS. The economic impacts of pollinator declines: an approach to assessing the consequences. **Conservation Ecology. v. 5, n.** 1, p. 8. 2001. [online] URL: http://www.consecol.org/vol5/issl/art8/

KEVAN, P. G. Honeybees for better apples and much higher yields: study shows pollination services pay dividends. **Canadian Fruitgrower.** v.14, n. 16, p., 1997.

KLEIN, A. et al. Importance of pollinators in changing landscapes for world crops. **Proceedings of the Royal Society.** v. 274, p. 303-313, 2007.

MATSUMOTO, S.; ABE A.; MAEJIMA, T. Foraging behaviour of Osmia cornifrons in an apple orchard. Scientia Horticulturae. v. 121, p. 73-79. 2009.

MATTU, V.K.; HEM, R.; THAKUR, M.L. Foraging behaviour of honeybees on apple crop and its variation with altitude in Shimla hills of western Himalaya, India. International Journal of Science and Nature. v.3, p. 296-301, 2012.

MCGREGOR, S. E. Insect pollination of cultivated crop plants. USDA, 1976.

NEILAND, R.; WILCOCK, C. Pollination failure in plants: why it happens and when it matters. TRENDS in plant Science, v.7, n. 6, p. 270-277, 2002.

PARANHOS, B. A. J. et al. Flight range of africanised honeybees, *Apis mellifera* L. 1758 (Hymenoptera: Apidae) in an apple grove. **Scientia agrícola, v** 54, n. 1-2, p. 85- 88,1997.

PETRI J. L. et al. Advances in Apple Growing in Brazil. **Rev. Bras. Frutic.,** Jaboticabal, Special Volume, p. 048-056. 2011.

RICKETTS, H.T.et al. Landscape effects on crop pollination services: are there general patterns? **Ecology Letters.** v. 11, p. 1-17, 2008.

ROHRER, J. R.; ROBERTSON, K. R.; PHIPPS, J. B. Floral morphology of Maloideae (Rosaceae) and its systematic relevance. **American Journal of Botany.** v. 81, p. 574- 581,1994.

SHEFFIELD, C. S.; SMITH, R. F.; KEVAN, P. G. Perfect syncarpy in apple *(Malus* x *domestica* 'Summerland McIntosh') and its implications for pollination, seed distribution and fruit production (Rosaceae: Maloideae). **Annalis of Botany.** v. 95, p. 583-591, 2005.

SHEFFIELD, C. S. et al. Winter management option for orchard pollinator *Osmia Lignaria Say* (Hymenoptera: Megachilidae) in Nova Scotia. **JESO.** v.139, p. 3-18, 2008.

SILVA, M. et al. Influence of *Trigona spinipes* Fabr. (Hymenoptera: Apidae) on the pollination of the yellow passion fruit. **Anais Sociedade Entomológica do Brasil,** v. 26, n. 2, p. 217-221, 1997.

SOSTER, M. T. B.; LATORRE, A. N. Evaluation of the phenology of the apple cultivars Imperatriz, Imperatriz Gala and Fuji in an orchard in Bom Retiro - SC. **Revista Biotemas.** v. 20, n. 4, p. 35-40, 2007.

TEPEDINO V. J. et al. Orchard pollination in Capitol Reef National Park, Utah, USA. Honey bees or native bees? **Biodiversity Conservation.** v.16, p.3083-3094, 2007.

THOMSON, J. D.; GOODELL, K. Pollen removal and deposition by honeybee and bumblebee visitors to apple and almond flowers. **Journal of Applied Ecology.** v. 38, p. 1032-1044, 2001.

VAISSIÈRE, B. E.; FREITAS, B.M.; GEMMILL-HERREN, B. Protocol do detect and assess pollination deficits in crops: a handbook for its use. **Food and Agriculture Organisation of the United Nations** (FAO/IFAD), Rome, 2010.

VICENS, N.; BOSCH, J. Pollinating efficacy of *Osmia comuta* and *Apis mellifera* (Hymenoptera: Megachilidae, Apidae) on "Red Delicious" Apple. **Environmental Entomology,** v. 29, p. 235-240, 2000.

WATSON, J. C.; WOLF, A. T.; ASCHER, J. S. Forested landscapes promote richness and abundance of native bees (Hymenoptera: Apoidea: Anthophila) in Wisconsin apple orchards. **Environ. Entomol.** v.40, n. 3, p. 621-632, 2011.

WEST, R. Pollination of apples by honey bees. State of New South Wales Department of Agriculture. Agnote DAI/132, Aug. 1999.

CHAPTER 2

Effect of densification with *Apis mellifera* L. hives/ha and pollen collector management on seed production in *Malus domestica* Borkh

Abstract: The intensive use of land by conventional agriculture reduces the richness and abundance of pollinators worldwide, causing a pollination deficit in pollinator-dependent crops, which represent around 70% of the world's cultivated plants. Because of this, the use of managed hives has become a common and efficient practice for promoting pollination in self-incompatible crops such as apple trees. However, the appropriate number of hives varies depending on the characteristics of the crop and its surroundings. The aim of this study was to assess the influence of the number and type of *Apis mellifera* hive management on the density of visits and seed production in an apple orchard in the Chapada Diamantina region of Bahia, Brazil. To this end, we comparatively assessed the density of bee visits and the number of seeds per fruit in a densified orchard with 7, 9 and 11 hives per hectare, and two types of hive management: with and without a pollen collector. Densification with *A. mellifera* (without a pollen collector) increased the density of visits (a = 0.05; $F_{2,45}$ = 22.29, p < 0.001) and the production of seeds per fruit (α = 0.05; $F_{2,27}$ = 4.0, p = 0.029) compared to the management practised by the producer, who used five unmanaged hives per hectare. However, densification with hives plus the pollen collector resulted in higher values for average density of visits and number of seeds per fruit than those obtained with densification without the pollen collector. To reduce the pollination deficit in the region's apple orchards, we recommend densities of at least 7 hives/ha plus a pollen collector and a centralised spatial arrangement in the orchard. Management should include prior acclimatisation of the hives, with 4-5 brood frames and feeding with an energy diet. Thus, improving management practices for A. *mellifera* hives and practices that are friendly to maintaining native pollinators in the orchard could reduce the high demand for hives to pollinate apple orchards in the region.

Keywords: Pollination deficit, bees, agroecosystem, Chapada Diamantina

INTRODUCTION

The intensification of agricultural production expands its borders into natural areas and this practice can lead to a reduction in ecosystem services such as pollination, due to changes in species richness and composition (Tilman *et al.,* 2002). The literature gathers evidence linking intensive land use by conventional agriculture to a reduction in the richness and abundance of pollinators around the world (Ricketts *et al.,* 2008; Viana et *al,* 2012). Habitat loss (Winfree *et al.,* 2009) and the use of agrochemicals (Freitas and Pinheiro, 2010) are factors that have a marked influence on bee populations, which are the main pollinators of many agricultural crops, especially in tropical countries such as Brazil (Imperatriz-Fonseca, 2004).

In a review covering crops from 200 countries, Klein *et al.* (2007) observed that the fruit, vegetable and seed production of 87 globally relevant crops depended on pollination by animals, while only 28 did not depend directly. According to the same authors, 60 per cent of world production comes from crops that do not depend on pollinators, 35 per cent depend on pollinators and 5 per cent are not assessed.

However, the scenario of low diversity of native pollinators, resulting from the homogenisation of landscapes around crops, causes pollination deficits in pollinator-dependent crops (Kremen *et al.,* 2002). Worldwide, techniques for managing pollinators in crops are concentrated on a few species, especially social bees of the genus *Apis* (Delaplane and Mayer, 2000) and some solitary species (Roubik, 1989; Bosch and Kemp, 2002; Imperatriz-Fonseca 2004). In Brazil, in addition to *Apis* management, some techniques have emerged as efficient in promoting the densification of orchards with solitary bees that nest in pre-existing cavities (Freitas and Oliveira-Filho, 2003) and meliponins (Venturieri, 2008). However, these practices are little used in Brazil, largely due to a lack of knowledge about basic aspects of biology, breeding techniques and taxonomic impediments (Imperatriz-Fonseca, 2012).

The use of managed *Apis mellifera* hives is a widespread and efficient practice for promoting pollination in self-incompatible crops such as apple trees, although the appropriate number of hives varies depending on the characteristics of the crop and its surroundings (Paranhos, 1998; Finta, 2004).

Delaplane and Mayer (2000) consider that the recommended density of hives depends on the attractiveness of the plantation, the population of unmanaged bees, the state of the *Apis* hives (i.e. strengthened by artificial feeding and improved for pollen collection), natural competition for nectar and other resources, the distance of the hives from the plants, weather conditions and the experience of the producer.

Studies have linked the use of pollen collectors to an increase in pollen collection activity by bees, due to a reduction in the amount of pollen grains stored in the hive (Nelson at al., 1987; Delaplane and Mayer, 2000). In almond and plum orchards, hives with pollen collectors had a higher number of pollen foragers than unmanaged hives (Webster *et al.,* 1985). Loper and Thorp (1984) reported that collectors remove around 16% of pollen at the entrance and can increase pollen collection by a factor of 1.8 in almond orchards, but can reduce brood comb construction when they collect 60% of pollen at the hive entrance (McLellan, 1974; Webster et al., 1985).

However, the efficiency of this management in hives used for pollination needs to be evaluated in a greater diversity of crops and agricultural landscape contexts to confirm the viability of management in cultivation since some types of collectors can cause the death of individuals (Wiser,

1992) and others allow the collection of different amounts of pollen (Loper and Thorp 1984), influencing the development of the hive, as well as the behaviour and foraging intensity of the bees in the orchard, thus generating inconsistent results.

In the Chapada Diamantina, apple productivity is 315 tonnes, with an average per hectare of only 7.5 tonnes, lower than that observed in other regions of the country where productivity is close to 40 tonnes/ha, with some orchards producing over 50 tonnes/ha (Petri *et al.,* 2011), even with the practice of densification with 5 hives/ha. This shows that the densification of orchards is not enough to improve the levels of pollination in the orchard. It is necessary to test higher densification gradients and improve hive management in order to intensify flower visits and increase seed production per fruit.

The selection of pollen collector management is based on previous observations that A. *mellifera* bees are more efficient at pollination when they collect pollen from apple tree flowers (Freire, 2009). Therefore, considering the local levels of pollination deficit, this study evaluated: (1) the effect of increasing levels of densification and (2) the management of hives with pollen collectors on seed production and the density of visits to apple tree flowers.

MATERIAL AND METHODS

Study area and orchard description

The study was carried out in a 43-hectare apple orchard belonging to the Bagisa S/A Agropecuária e Comércio company, located in Cascavel (13°24'50, 7"S and 41°17'7.4"O), a district of Ibicoara in the Chapada Diamantina, state of Bahia (Figure 1) during the 2011 production cycle.The region has predominantly Cerrado vegetation, an altitude of approximately 1100 metres and an average annual temperature of 21°C, with average highs of 26°C and average lows of 16°C. The rainy season runs from November to March, with annual rainfall of 757 mm (Alves, 2009).

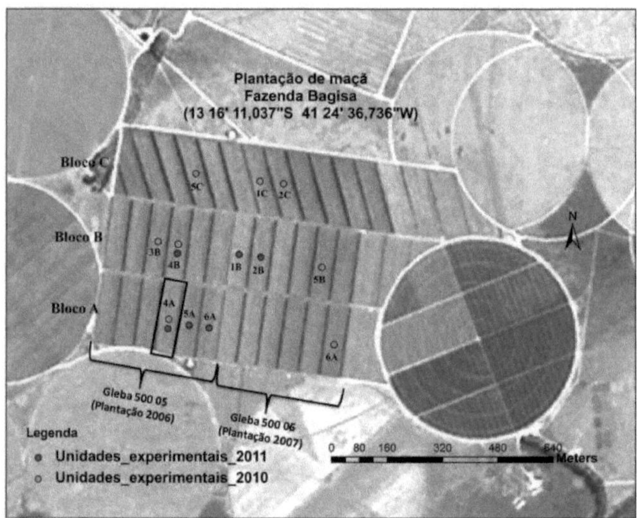

Figure 1: Top overview of the apple orchard (43 ha) at Fazenda Bagisa, Cascavel district, Ibicoara, Bahia, Brazil. The green circles (1C, 2C, 5B, 6A, 5C, 4B, 3B) indicate the experimental plots used to study the pollination deficit in 2010 and the red circles in 2011. The plots were selected by lottery for the delimitation of the 50 x 25 m sampling units. Supervised high-resolution satellite image (SPOT) (pixels 5 m on a side) taken in September 2008 of the Cascavel district, Ibicoara, Bahia, Brazil. Image taken in September 2008.

In the region, planting of the commercial variety Eva (a cross between "Gala" and "Ana") (IAPAR, 2009) began in 2005. It differs from the Gala variety in its low chilling requirements (between 300 and 350 hours) for breaking dormancy and growth of the vegetative branches, early ripening and larger fruit size. The fruits are striated red on a whitish-yellow background, with a conical to rounded shape. The Princesa cultivar, developed by Epagri in Caçador-SC, is used exclusively as a pollen donor because it has a low fruit yield but a high volume of flowers with prolonged flowering, producing a large quantity of pollen.

The orchard is divided into 36 plots of 1.2 hectares each, containing 14 rows spaced 4 metres apart, with 5 Evas to 1 Princesa (5:1 ratio). The latter accounts for around 10-12% of the orchard (Figure 2).

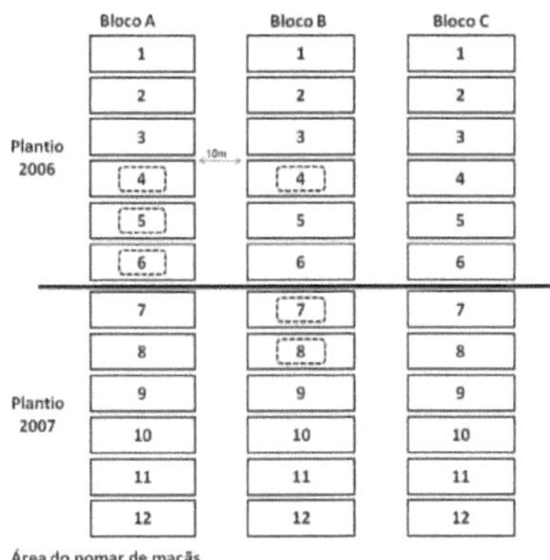

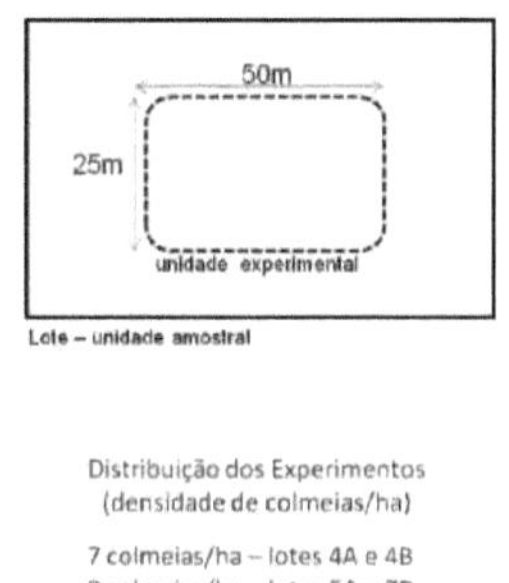

Figure 2: On the left, the 36 plots of the Eva and Princesa apple orchard in Chapada Diamantina, BA, are shown schematically. Of these, 18 plots were planted in 2006 and 18 plots were planted in 2007. The distance between the blocks is 10m and each block (A, B and C) contains 12 plots of 1.2 hectares (60 x 200m) each. The experimental units are rectangular areas of 25 x 50m, equal to 1250 m^2). The numbering of the plots drawn, where the experimental units were delimited, are surrounded by a rectangle in a dashed line. In the top right-hand corner, the figure illustrates the relative position and size of the experimental unit (dotted line) within the plot (solid line) on which the densifications were tested in 2011.

Flowering takes place in July by inducing dormancy. To do this, the grower removes the mature foliage after the harvest, preventing the buds from entering endo-dormancy, using a mixture of 0.75 TA 35 + zinc + 11.25 copper sulphate. To induce flowering, 1500 litres of a mixture consisting of 12 dormex (hydrogen cyanamide), 0.75 pyrimex, 60 Agrex oil are applied to each plot (1.2ha). The flowering inducer is applied during the month of June, and in 2011 the schedule for inducing flowering in the plots was carried out by the producer in six stages. Thus, at each interval the inducer was applied to six plots (2 plots in each block), sequentially until all the plots were induced to flower.

Sampling procedure

The procedures adopted in this study for delimiting the experimental units and quantifying the number of seeds per fruit, flower density, visitor density and diversity are based on the manual: "Protocol to Detect and Assess Pollination Deficit in Crops - FAO/IFAD" (Vaissière *et al.*, 2010, 2011).

The pollination deficit was obtained from the results of the deficit assessment carried out in 2010 using 5 hives/ha in 8 experimental units (Figurei). In each of the eight experimental units selected, 16 plants were chosen and distributed into 4 plots, each consisting of 2 Eva and 2 Princesas, totalling 8 plants of each cultivar. On each individual plant, 2 buds ("King Blossom", which corresponds to the central flower of the inflorescence) were selected in similar positions, but in different bouquets, which were marked and bagged until anthesis, one bud for each treatment, namely: flower open to floral visitors and not manipulated for natural pollination (control) and flower for manual cross-pollination.

For cross-pollination, dehiscent anthers from Princesa flowers were rubbed lightly against the receptive stigmas of Eva flowers, observing the deposition of pollen grains on the stigmas using a magnifying glass (lOx). The naturally pollinated flowers were re-bagged after 24 hours and the hand-pollinated flowers were re-bagged immediately after pollen transfer. Both remained bagged for 90 days, when they were collected to count the number of seeds.

The experiments with 7, 9 and 11 hives per hectare and management of *Apis mellifera* hives with or without a pollen collector to test the influence on seed production and the density of flower visits were carried out in the six experimental units delimited for this study by drawing lots (Figure 2). The choice of hive densities, with a minimum of 7 hives/ha, was based on the following factors

The densities were tested in the following order: 7 hives/ha, 9 hives/ha and 11 hives/ha, with the aim of standardising the densities in view of the progressive increase in the number of flowering plots in the orchard, provided for in the producer's induction schedule in 2011. Thus, the number of hives in the orchard increased: 84 (7 hives/ha - 12 flowering plots), 162 hives (9 hives/ha - 18 flowering plots) and finally 330 hives (11 hives/ha - 36 flowering plots) (Figure 2).

The hives were installed in the orchards at night, 24 hours before the start of the experiments, in order to acclimatise them. Each densification level was tested in two experimental units for two consecutive days. On the first day the hives were kept without a pollen collector and on the second day the pollen collector was added to the hive entrance (Figure 3). The pollen collectors consist of two parts: a plastic screen with 4.5mm perforations to scrape the pollen from the corbiculae of the bees' hind legs when they pass through the holes to enter the hive; the other part is a screened collection tray to prevent the bees or ants from removing the pollen.

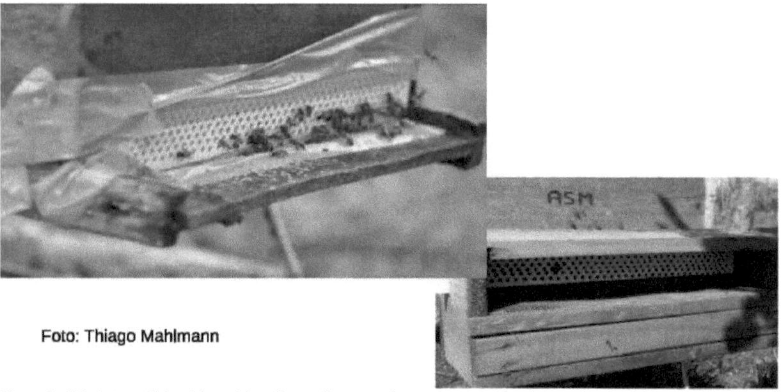

Foto: Thiago Mahlmann

Figure 3: (A) *Apis mellifera* hive with pollen collector at the entrance; on the right, a detail of the pollen collector installed in the hive in Bagisa's apple orchard in 2011. Chapada Diamantina, BA.

Management of beehives used in orchard densification

The hives used in the experiment were transported during the month of May 2011 from the municipality of Andaraí to the region of Bonito (located in the Chapada Diamantina region, BA), whose climatic conditions and altitude are similar to those of Cascavel (the site of the field experiments). This procedure was adopted to acclimatise and strengthen the hives, since in this region there was a large quantity of native plants in bloom and the resources provided by a eucalyptus plantation next to where the hives were placed were also used.

In order to equalise the hives supplied by the beekeeper, they were inspected taking into account the external state of conservation of the boxes, the replacement of old combs, queen replacement (all the queens in the hives had been replaced the previous season, i.e. they were less than 1 year old), bee diseases, the size of the population, the increase in brood areas and the presence of 3 - 4 frames of brood (moderately weak hive). This number of frames of brood is considered low in relation to the recommended minimum, since strong hives have a high number of adult bees and a number of frames of brood between 6-10 per hive, guaranteeing an adequate number of workers, eggs, larvae and young pupae for a future population of foragers (Delaney and Tarpy, 2008).

The hives were placed in the centre of the plots along the rows, 3 metres apart, during peak flowering and kept in the field for 10-15 days until the end of flowering. The rational Langstroph boxes, made from pine wood, were kept on metal supports to prevent ant attacks.

The selection of pollen collector management was based on empirical work that relates the inclusion of *a pollen* collector at the entrance to the hives to an increase in the number of visits by *A.mellifera to the* orchard flowers, since it prevents the bees from entering with a full load of pollen, removing it from the hairs and corbiculae as they try to pass through the circular openings in the screen, falling by gravity into the drawer (Figure 3).

Effects of densification and management of *Apis mellifera* hives

To analyse the effect of densification with 7, 9 and 11 hives/ha on the number of seeds per fruit, the density of visits and the diversity of floral visitors to the apple tree, individual plants were marked in each experimental unit. The density of flowers on Eva and Princesa was quantified in the six experimental units, densified with *Apis mellifera* hives, with and without a pollen collector, as this factor influences the attractiveness and visitation of the bees to the flowers.

The procedure for measuring the number of seeds per fruit, flower density, visitor density and visitor diversity was carried out on different plants; only flower density and visitors were quantified on the same plants, whose number and spatial distribution in the experimental units were defined according to the protocol for studies on pollination in apple orchards (Vaissière *et al.*, 2011). Four points were selected for each parameter assessed in the experimental units (Figure 4), with the exception of visitor diversity, which was assessed at six points (Figure 5). Each point is made up of 4 plants: 2 individuals of the commercial variety (n=8 eva) and 2 individuals of the pollen donor variety (n=8 princesa). To estimate the diversity of visitors, five points were used with pairs of evas (n=10 evas) and one pair of princesses (n=2 princesses).

To assess the effect of each hive density tested (7, 9 and 11 hives/ha) on the number of seeds in the apple fruit, 16 plants of the commercial variety (eva) were marked, 8 plants for the test with hives without a pollen collector and 8 plants for the test with a pollen collector.)

Thus, 32 flowers were marked in each experimental unit (4 flowers/plant/experimental unit), 16 of which were exposed to visitation during the test with hives without a pollen collector (24-hour period) and 16 flowers were exposed to pollinators the following day during the tests with hives plus the pollen collector (24-hour period). After this period, the marked flowers were bagged until the initial fruit formation to count the number of seeds.

The pollination deficit values obtained in 2010 are used as a reference for evaluating the effect of the densifications carried out in this study. To do this, the results of the tests were compared with the production of seeds obtained by hand pollination and natural pollination in 2010, which were obtained under the same experimental design described in a previous version of this manual (Vaissière *et al.*, 2010).

The flower density of the eva and princesa varieties was recorded once in each experimental unit (n= 8 eva /8 princesa) at 4 points marked in rows in the centre of the experimental area (Figure 4). The flowers on each individual were quantified on 2 representative opposite branches on the plant (base width between 4-5 cm). Flower production was estimated by multiplying the average number of flowers by the total number of branches on the plant. In total, 48 plants of each variety (n= 96 plants) were sampled in the six experimental units.

Visitor density was quantified at the same points used to assess flower density (Figure 4). The records were made by a collector travelling through the rows in the morning (beginning at 1pm) and afternoon (beginning at 2pm), for two consecutive days, during the orchard's most intense flowering phase. Using two counters, the visits were recorded immediately on a sample of 50 flowers on each individual, this number being defined according to the availability of flowers on the plants. This assessment was carried out in good weather conditions, with a temperature of 15°C or above, low wind speed, no rain and dry vegetation. Data was collected at the start of the flowering period, i.e. when 10% or more of the plants had started flowering and had flowers at anthesis to allow the bees to forage (Vaissière *et al.*, 2011).

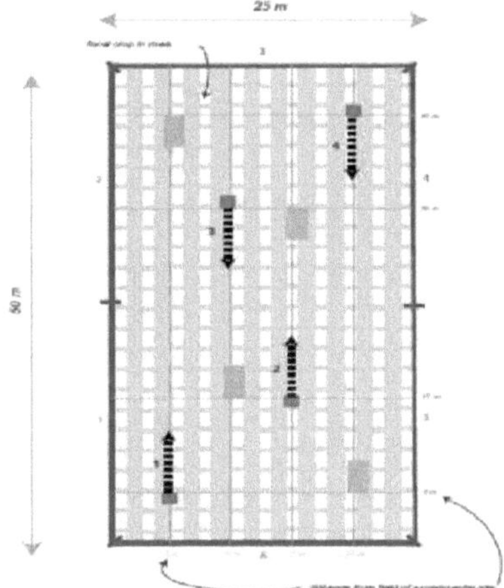

Figura 4: Diagram of the experimental unit for assessing flower density and visitor density in apple trees (yellow rectangles indicate the location of the 4 sampling points, each with *2 evas: 2 princesses*). Figure taken from the *Protocol to Detect and Assess Pollination Deficit in Crops - FAO/IFAD"* (Vaissière *et al., 2011*).

The six points (n=12 plants) used to sample the diversity of visitors were selected on the edges of the experimental area, so that the collector circled the points, travelling around the experimental area (Figure 5). Specimens were collected using an entomological net for 5 minutes at each point, totalling 30 minutes per experimental area. The specimens captured were killed in a death chamber containing ethyl acetate and then transferred to vials labelled with the date of capture, the name of the experimental area and the collector. They were then kept under refrigeration (-20°C), subjected to the mounting steps and identified.

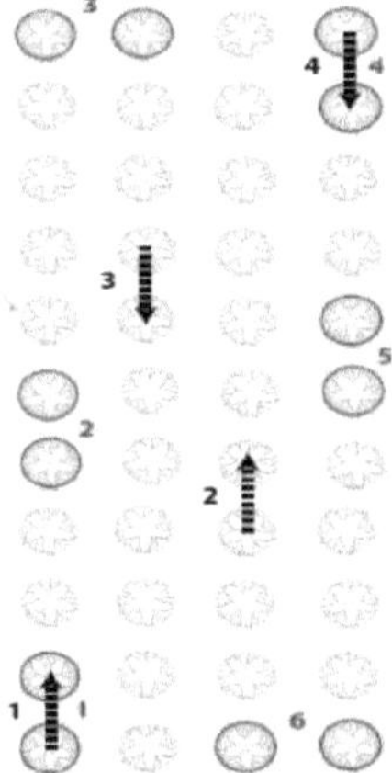

Figura 5: Diagram of the experimental unit for sampling the diversity of floral visitors in apple trees. The circles indicate the apple trees; the black arrows indicate the plants used to assess the density of visitors and the red circles indicate the plants used to assess the diversity of floral visitors. Taken from *Protocol to Detect and Assess Pollination Deficit in Crops - FAO/IFAD "* (Vaissière *et al.,*

2011).

Statistical analysis

To test the hypothesis that densification influences the density of floral visitors, we used the One Way ANOVA test (in the treatment without a pollen collector) with Tukey's multiple comparisons test to detect where the difference lay. To test the hypothesis that density influences the total number of seeds produced per fruit (10 seeds maximum per fruit), we used the one-way ANOVA test with three levels.

To compare the density of floral visitors between the different densities in the pollen collector treatment, we used the non-parametric Kruskal-Wallis test, since the data set showed heteroscedasticity. Dunn's test (with automatic α adjustment) for multiple comparisons was used a posteriori to find out which categories were different. All the analyses were carried out using *SPSS 19.0 for Windows* software.

RESULTS

In the densities tested, the average number of seeds in the fruit varied between 5-8 (Figure 6A-B). The increase was progressive and significantly different between the treatments without a pollen collector ($\alpha = 0.05$; $F_{2,27} = 4.0$, $p = 0.029$), obtaining fruit with eight seeds on average in the densification with 11 hives/ha (Figure 6A). The addition of the pollen collector to the hives did not significantly influence the average number of seeds in the fruit ($\alpha = 0.05$; $F_{2,27} = 0.18$, $p = 0.83$) in the densities, but exceeded the average number of seeds in the fruit in the densities with 7 hives/ha without a pollen collector (Figure 6B).

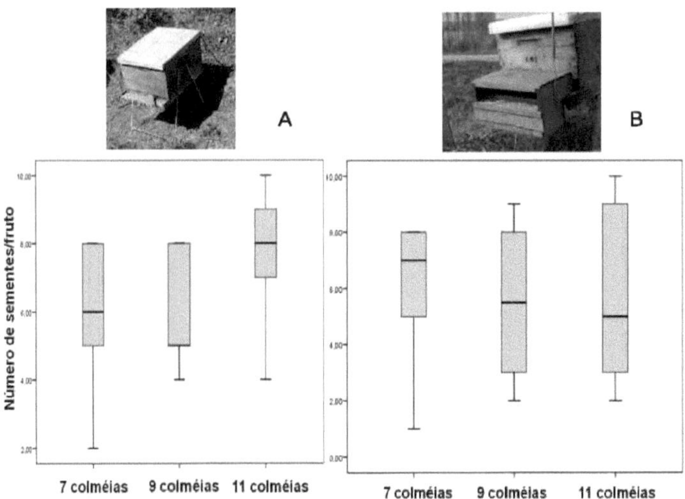

Figura 6:   Seed production per fruit of the Eva cultivar in relation to densification levels with unmanaged *Apis mellifera* hives (A) and managed with a pollen collector (B) in Chapada Diamantina, BA.

The densification of the orchard with 7 and 9 hives/ha without a pollen collector generated a similar increase (between 5 and 10%) in the density of floral visitors to the Eva variety (Figure 7), which was significantly higher only in the densification with 11 hives/ha ($\alpha = 0.05$; $F_{2,45} = 22.29$, p < 0.001) (Figure 7). When added with a pollen collector, densification with 7 and 9 hives/ha resulted in different values, with a tendency towards a progressive increase in the density of visitors. However, densification with 11 managed hives/ha resulted in lower visitor density values than expected ($\alpha = 0.05$; KW = 31.79, p < 0.001), compared to the other densifications. The flower density of the pollen donor variety was higher than that of the Eva variety ($\alpha = 0.005$; KW = 20.60; p<0.001) (Figure 8). The number of Apis mellifera visits was comparatively lower in eva than in princesa (Figure 9A and 9B). For both cultivars, densification with nine hives managed with a pollen collector resulted in a higher number of *Apis mellifera* visits per minute, with the Princess cultivar receiving more than double the number of visits recorded on the Eva cultivar plants (Figure 9).

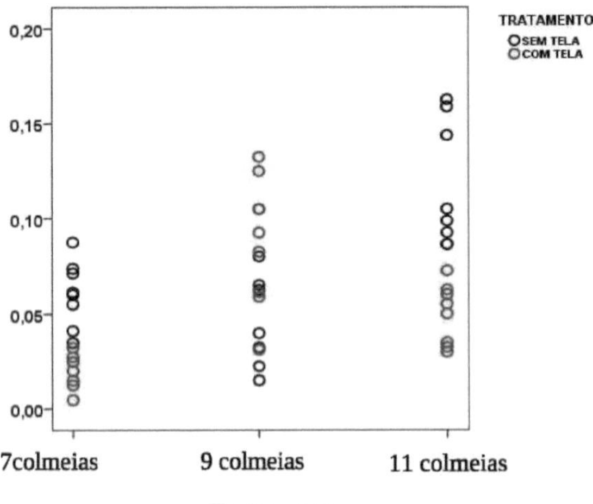

Tratamento

Figura 7: Relationship between densification levels with unmanaged *Apis mellifera* hives (black circles) and those managed with a pollen collector (red circles) and the density of visits (number of visits/50 flowers/plant) to the flowers of the Eva cultivar in the orchard in Chapada Diamantina, BA, 2011. The points on the graph represent the plants (8 marked plants in each experimental unit per treatment) sampled.

Number of *Apis mellifera visits/plant* E va/ min   Number of *Apis mellifera visits/pla* nta Princesa/min mmmimin

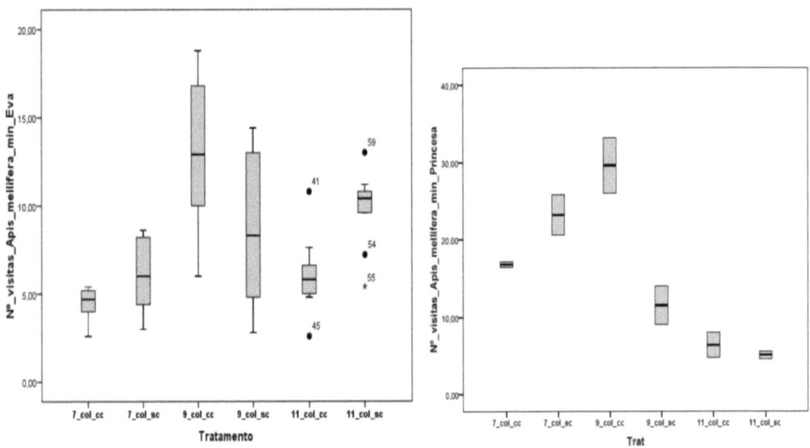

Figura 8: Variation in flower density in the Eva (commercial) and Princesa (pollinator) apple varieties in Chapada Diamantina, BA. Flower density was measured only once when the density of flowers in the sampled plot was at peak flowering (visual estimation) in 2011.

Figura 9: Number of *Apis mellifera* visits per plant per min[1] in the Eva (n=10) (A) and Princesa (n=2) (B) varieties - at each densification level (7.9 and 11 hives/ha) and type of hive management (cc - with pollen collector, sc - without pollen collector) in the Chapada Diamantina orchard, BA. The filled bars indicate the mean value and the vertical lines the standard deviation from the mean. Year 2011.

Eighteen species of flower visitors represented by the orders Diptera (llsp.), Hymenoptera (6sp.)

and Lepidoptera (2sp.) were collected in the apple orchard (Table 1).

Table 1: Diversity of flower-visiting insects collected with an entomological net in an apple orchard in 2011 in Chapada Diamantina, BA. Key: s/c = hives without pollen collector; c/c = hives with pollen collector.

Order	Spit	Number of individuals collected												Total individuals
		11 hives/ha		11 hives/ha		9 hives/ha		9 hives/ha		7 hives/ha		7 hives/ha		
		s/c	c/c	s/c	c/c	s/c	c/c	s/c	c/c	s/c	c/c	s/c	c/c	
Hymenoptera	*Apis mellifera* Linnaeus, 1758	261	319	463	151	397	447	539	555	108	153	247	133	3773
	Trigona spinipes Fabridus, 1793			1			1	2						4
	Halictidade sp.1			1										1
	Exomalopsis (Exomalopsis) analysed Spinola, 1853	1		1								1		3
	Dialictus sp			1										1
	Zethus sp								1					1
Diptera	Anthomyiidae sp.1							1						1
	Musddae sp.1	1						2			1			4
	Musddae sp *2*		1						4					5
	Chrysomelidae sp.1	1												1
	Chrysomelidae sp.2				1									1
	Syrphidae sp.2												1	1
	Syrphidae sp.4							8						8
	Syrphidae sp.6						5						1	6
	Syrphidae sp.1							1						1
	Syrphidae sp.5		2	5										7
	Syrphidae sp.9		1			2			2	2				7
Lepidoptera	Lepidoptera sp 1	1						1		1		1		4
	Hesperidae sp 1					1								1
	Number of species	5	4	6	2	3	3	7	4	3	3	2	3	

The distribution and numerical representation of species (8, 13 and 13 species with 7, 9 and 11 hives/ha, respectively) varied between densities and management, but in general the presence of the pollen collector coincided with a lower diversity of visitors, especially bees (Tableai).

DISCUSSION

The increase in the density of hives tested in this study increased seed formation compared to the use of 5 hives/ha practised by the producer in 2010, when the fruit reached an average of 4-5 seeds per fruit (chapter 1, in this dissertation). The addition of pollen collectors reduced the number of hives/ha needed to obtain 7 seeds per fruit, compared to densification without pollen collectors. We therefore believe that pollen collector management was more efficient and economically viable for reducing the pollination deficit locally.

However, the number of hives/ha needed to reduce the pollination deficit in Chapada Diamantina exceeds the recommended density for apple orchards studied, which ranges from 1 to 4 hives/ha (Keogh *et al.,* 2008, Delaney and Tarpy, 2008, Aichele et *al.,* 2009, Khan *et al.,* 2004). In Brazil, Paranhos *et al.* (1998) suggest using 2.5 hives/ha for the Anna cultivar and, in Minas Gerais, 0.5 hives/ha is enough for efficient pollination of the Eva cultivar (Communication from the producer, José Lázaro).Despite this, we have seen an increase in seed production, as predicted by Kevan (1997) for apple orchards in Ontario, who showed that the addition of one *A.mellifera* hive per hectare resulted in the production of one more seed per fruit and an increase in productivity when compared to an orchard without bees.

In the area studied, the addition of hives in 2011 increased productivity by 25%, from 7.5 tonnes/ha in 2010 to approximately 20 tonnes/ha in 2011. However, this figure only represents the productivity obtained by Fazenda Corredor, in Piedade do Rio Grande, MG, which reaches 40ton/ha (IBGE data for 2010), for the Eva cultivar. According to the theoretical model of Degrandi-Hoffman *et al.* (1987), increasing the number of beehives per hectare resulted in greater fruit set, but there was a point at which the number of fruits formed began to decrease, which made the use of more than 5 hives/ha unnecessary.

A. mellifera predominated as a flower visitor in all treatments and the intensification of densification, with hives of this species, from 5 to 7 hives/ha increased the number of visits by these bees to the flowers of the Eva cultivar, as evidenced for other crops (Dupree *et al.* 1999). The high number of visits by A. *mellifera* to the pollen donor cultivar (average = 18 visits/plant/min-1) with the use of 7 hives/ha plus the pollen collector, driven by the flowering density of this variety, increases the chances of transferring compatible pollen to the stigmas of the commercial variety so that 6 and 8 visits/plant/min may be enough to increase pollination efficiency in the orchard studied.

Using a pollen collector at 9 hives/ha increases the visit rate to levels similar to those estimated for the "Delicious" apple variety, between 20 and 25 visits/plant/min (Mayer *et aZ.*, 1986). The observed reduction in the number of visits with 11 hives/ha may indicate that (1) above 9 hives/ha there is a negative effect of densification or (2) that the time taken for the hives to adapt to the pollen collectors was insufficient, preventing more champion bees from leaving (Free 1967, Goodwin *et al.* 1984), resulting in a reduced density of visitors.

Studies testing pollen collector management have shown an increase in the amount of pollen collected by the bees of up to 80% (Rashad and Parker 1985) and in the amount of pollen adhered to the body of the bees leaving the hive (Manning *et al.,* 2010). These authors found that the presence of the collector produces an increase of between 278-352% in the amount of pollen grains on the body of the bees leaving the nest, compared to hives without this management, which are normally used for pollination.

The flower morphology of the Eva cultivar allows nectar to be collected without contacting the reproductive structures (chap 1, in this dissertation), similarly to other apple cultivars (Schneider et al., 2004), so that the efficiency of bees in pollination is greater during visits to collect pollen. Thus, this technique intensifies the activity of the visitor and favours the efficiency of the pollinator, increasing the chances of adequate pollination. This hypothesis is reinforced by the positive relationship found between seed production and the density of visits by bees from hives with pollen collectors.

In this way, the stimulation of pollen collection in flowers promoted by the use of pollen collectors in hives favours pollination for two reasons: (1) it directs the visitor's landing and movement on the anthers and stigmas; (2) it increases the chances of pollination occurring during the ovule's longevity period, which in apples usually lasts between 1 and 2 days (Sapir and Goldway, 2007), since this resource is sought in young flowers.

The literature on the management of *A.mellifera* hives for pollination is still scarce and controversial. Most authors adopt strong hives for use in pollination, since weak hives have fewer nurse bees (Goodwin *et al.,* 1984). Thus, in hives used for pollination, the number of brood frames suggested varies from 4 to 6 (Mayer *et al.,* 1986; Hoopingarner and Waller, 1993) and 6 to 10 brood frames (Ben-Porat et al., 1997, Delaplane *et al.,* 2000, Dag et *al.,* 1999, 2003).Pollination services in crops isolated from natural areas, as is the case in the region studied, obtained through densification with hives depend on proper management. One factor that may be contributing to the high demand for bees is the use of intermediate or weak hives, as evidenced by the number of brood frames between 3-4, based on the premise that developing hives collect more pollen to provide for their brood.

In the Chapada Diamantina, the intensification of agricultural activity, the low percentage of available habitat (23%) within a radius of 2 km from the orchard, and conventional agricultural practices, which do not include pollinators, such as the use of pesticides, must contribute to the absence of other bee species in the orchard studied (Kremen *et al.,* 2002; Kevan *et al.,* 2002; Foley, 2005). According to Watson *et al.* (2011), the abundance of native pollinators decreases in plantations that are isolated from natural areas, where there is a lack of natural nesting habitat and alternative food resources, and are therefore considered important ecological resources for maintaining pollination services.

ACKNOWLEDGEMENTS

The authors would like to thank CNPq and GEF/FAO/FUNBIO/UNEP for their financial support, which made the research possible. To Professors Guido Castagnino (UFBA) and Kátia Gramacho (UNIT) for their guidance on the management of Apis *mellifera* hives and for their collaboration in the field. The company Bagisa S/A Agropecuária e Comércio for its logistical support and access to the orchards. Blandina F. Viana thanks the CNPq research productivity grant. Fabiana O. da Silva thanks CAPES for the PNPD scholarship; Antonio da Costa Diakos thanks FAO/FUNBIO for the master's scholarship.

REFERENCES

AICHELE, T.; SHANE, B. 2009. The fire blight Interactive predictor. Crop Advisory Team. Michigan State University Extension, v 24, n. 4, May, 2009.

BEN-PORAT, A.; DORON , L; DAG, A. Apple pollination. **Alon Hanotea,** v. 51,n. 1, p. 76- 78 1997.

BOSCH , J.; KEMP, W. P. Developing and establishing bee species as crop pollinators: the example of Osmia spp. (Hymenoptera; Megachilidae) and fruit trees. **Bulletin of Entomological Research,** v. 92, p. 3-16. 2002.

DAG, A.; DORON, L; STERN, R.A. Recommendation for pollination of apple, pear and plum. Newsletter - The Extension Service, Galilee-Golan Region (in Hebrew), 2 p. 2003.

DEGRANDI-HOFFMAN, G.; HOOPINGARNER, R.; PULCER, R. Redapol: pollination and fruit-set prediction model for "delicious" apples. Enviromental Entomology, v.16, p.309-318,1987.

DELANEY, D.; TARPY, D. 2008. The role honey bees in apple pollination. Department of Entomology, North Carolina State University, NC State Apiculture Programme. Beekeeping note 3.03. Available at: www.cals.ncsu.edu/entomology/apiculture/pdfZ3.03copy.pdf. Accessed on: 14 Aug.2012.

DELAPLANE, K. S.; MAYER.; D. F. Crop pollination by bees. CAB Publishing, Walling Oxon Ox 108 DE UK, UK the University Press, Cambridge, 362p. 2000.

DUPREE, C. S et al. A Guide to: Managing bees for crop pollination. Canadian Association of Professional Beekeepers, p. 41, 1999.

FINTA, K. **Insect pollination of apple orchards.** University of West Hungary, Faculty of Agricultural and Food Science, 2004.

FOLEY, J. A. et al. Global consequence of land use. American Association for the Advancement of Science, New York, v. 309, p. 570-574, 2005.

FREITAS, B. M.; PINHEIRO, J. N. Sub-lethal effects of agricultural pesticides and their impacts on pollinator management in Brazilian agroecosystems. **Oecologia Australis,** v. 14, p. 282-298, 2010.

GOODWIN, R. M.; PERRY, J. H. Use of pollen trap to investigate the foraging behaviour of honey bee colonies in kiwifruit orchards. Ruakura Apicultural Research an Advisory Unit MAF Technology, 1984.

GOTELLI, N. J.; ELLISON, A. M. Principles of statistics in ecology. Artmed Editora, p. 307-366, 2004.

HOOPINGARNER, R. A.; WALLER G. D. 1993. Crop pollination. In: GRAHAM, J. M (Ed.) The Hive and the Honeybee, Dadant and Sons, Carthage, Illinois, USA, p.1044-1080.

IMPERATRIZ-FONSECA, V. L.; CANHOS, D. A. L.; ALVES, D. A.; SARAIVA, A. M. (Orgs.). Pollinators in Brazil: contributions and perspectives for biodiversity, sustainable use, conservation and environmental services. São Paulo: Editora da Universidade de São Paulo, 488p, 2012.

AGRONOMIC INSTITUTE OF PARANÁ. Available at: www.iapar.br > Accessed on: 20 January 2009.

KEOGH, R et al. Apples. Australian Government, Rural Industries Research and Development Corporation Publication No. 10/109, Horticulture Australia Limited, 2008.

KEVAN, P.G. Honeybees for better apples and much higher yields: study shows pollination services pay dividends. Canadian Fruitgrower, v. 14, n. 16.1997.

KEVAN, P. G.; RICHARDS, K. W. Aspects of bee diversity, crop pollination, and conservation in Canada. The Conservation Link Between Agriculture and Nature - Ministiy of Environment/Brasília, 2002.

KHAN, M.; KHAN, M. The role of honey bees *Apis mellifera* L. (Hymenoptera Apidae) in pollination of apple. **Journal of Biological Science,** v.3, p. 359-362. 2004.

KLEIN, A. M. et al. Importance of pollinators in changing landscapes for world crops. Royal Society of Byology, v. 274, p. 303-313. 2007.

KREMEN, C.; WILLIAMS, N. M.; THORP, R. W. Crop pollination from native bees at risk from agricultural intensification. Stanford University, Stanford, CA. **PNAS,** v. 99, p. 16812-16816, 2004.

LOPER, G. M.; THORP, R. W.; BERDEL, R. Improving honey bee pollination efficiency in almonds. **California Agriculture,** v. 39, n. 11-12, p. 19-20,1985.

MANNING, R.; SAKAI, H.; EATON, L. Methods and modifications to enhance the abundance of pollen on forager honey bees (Apis mellifera L.) exiting from beehives: implications for contract pollination Services. **Australian Journal of Entomology,** v. 49, p. 278-285, 2010.

MAYER, D. F.; JOHANSEN, C. A.; BURGETT, D. M. 1986. Bee pollination of tree fruits. Pacific Northwest Extension Publication, PNW 0282.

MCLELLAN, A. R. Some effects of pollen traps on colonies of honeybees. **Journal of Apicultural Research,** v. 13, p. 143-148,1974.

NELSON, D. L.; MC KENNA, D.; ZUMWALT, E. The effect of continuous pollen trapping on sealed brood, honey production and gross income in Northern Alberta. Apicultural Research, 1987.

PARANHOS, B. A. J.; WALDER, J. M. M.; MARCHINI, L. C. Density of hives of Africanised bees, Apis mellifera 1. 1758 (hymenoptera: apidae), for pollinating apple cv. anna **Scientia agrícola,** v. 55, n. 3,1998 .

PETRI J. ET AL. Advances in apple growing in Brazil. Ver. Bras. Frutic., Jaboticabal, special volume, p.48-56. 2011.

RICKETTS, H. T. et al. Landscape effects on crop pollination services: are there general patterns? **Ecology Letters,** v. 11, n. 5, p. 499-515, 2008.

ROUBIK, D. W. Foraging Behaviour of competing africanized honeybees and stingless Bees. **Ecology,** v. 61, n. 4,1989.

SCHNEIDER, D.et al. A comparative study of the superior fertility of 'Smoothy Golden Delicious' apple. **Journal of Horticultural Science and Biotechnology,** v. 79, p. 596- 601, 2004.

TILMAN, D. et al. Agricultura! sustainability and intensive production practice. Department of Ecology, Evolution and Behavior. University of Minnesota, 2002.

VAISSIÈRE, B. E.; FREITAS, B. M, GEMMILL-HERREN, B. Protocol to detect and assess pollination deficits in crops: a handbook for its use. Food and Agriculture Organisation of the United Nations (FAO/IFAD), Rome, p.70. 2011.

VIANA, B. F. et al. Pollination in the context of the landscape: what we really know and what we need to know. InPERATRIZ-FONSECA, V. L.; CANHOS, D. A. L.; ALVES, D. A.; SARAIVA, A. M. (Orgs.). Pollinators in Brazil: contributions and perspectives for biodiversity, sustainable use, conservation and environmental services. São Paulo: Editora da Universidade de São Paulo, 488p, 2012.

WEBSTER et al. Effects of pollen traps on honey bee (Hymenoptera: Apidae) foraging and brood rearing during almond and prune pollination. **Environmental Entomology** 14(6): 686-686, 1985.
WINFREE, R et. al. A meta-analysis of bees responses to anthropogenic disturbance. Ecological Society of America, v. 90, n. 8, p.2068-2076, 2009.

FINAL CONSIDERATIONS

In general, the use of high levels of densification with hives/ha, such as those used in the area studied, are related to environmental factors and various landscape factors. The low percentage of available habitat (23 per cent) around the orchard reduces the diversity and support capacity of natural bee nests. The proximity of natural areas to the orchard reduces the demand for bees in orchards and, consequently, the costs of this input. In general, there is no evidence of the effect of densification or management on the diversity of native bees recorded in the orchard studied, which occurred in low density.

Densification with *Apis* bees promotes greater seed production per fruit, thus reducing the pollination deficit in the orchard studied. The data suggests that the densification levels needed to alleviate the pollination deficit and increase productivity are becoming impractical. On the other hand, it must be considered that a combination of factors such as the lack of attention paid to hive management, especially the use of weak hives; the difficulty in synchronising flowering in the pollinator and producer varieties due to induction; and the low diversity of native bees can together compromise foraging and pollen transfer between varieties.

Therefore, in order to make efficient use of the densification technique, it is necessary to invest in improving hive management techniques to increase their efficiency and, above all, in making some conventional agricultural practices more flexible in order to provide suitable habitats for bees in and around the orchard. Pollination should be incorporated into the management system of the apple orchard, not just as an input but as an integral part of the management system.